MILITARY AIRCRAFT

Published by:

Airlife Publishing Ltd
101 Longden Road
Shrewsbury SY3 9EB
England
Telephone: 0743 235651
Fax: 0743 232944

Produced by Aerospace Publishing Ltd and
published jointly with Airlife Publishing Ltd

© Aerospace Publishing Ltd 1994

First published 1994

ISBN 1 85310 537 6

Printed in Singapore

THE VITAL GUIDE TO

MILITARY AIRCRAFT

EDITOR: SOPHEARITH MOENG

Airlife

England

Aermacchi M.B.339

The Italian air force's 'Frecce Tricolori' aerobatic team operates the M.B.339PAN variant. This has a smoke-generating pod, but not the tip tanks common to most M.B.339s.

From experience gained by nearly 800 of its successful MB-326 jet trainers, Aermacchi developed a replacement version as the **M.B.339**. This retained the MB-326's licence-built Viper 632-43 turbojet and basic airframe aft of the rear cockpit but introduced pressurisation in a revised, deeper forward fuselage, an extended canopy with improved all-round view, a larger fin and more advanced avionics. Two prototypes preceded Italian air force (AMI) orders for 100 production **M.B.339A** trainers, with the first initially flying on 12 August 1976.

Service and operators
The M.B.339 entered AMI service in August 1979. Italian-developed converted variants include 19 **M.B.339PAN**s for the 'Frecce Tricolori' aerobatic team and eight specially-equipped **M.B.339RM**s for radio calibration duties. Export customers for the M.B.339A comprise the Argentine navy (10), Dubai air wing (five), and the air forces of Ghana (two), Malaysia (13), Nigeria (12), and Peru (16).

Attack variants
The basic trainer has been developed with enhanced ground-attack capability. A single-seat variant was built in 1980 as the **M.B.339K Veltro 2** which received no orders. The improved **M.B.339B** trainer introduced the uprated 19.57-kN (4,400-lb st) Viper 680 and enlarged wingtip tanks with greater capacity. None have been sold. The **M.B.339C** was developed as a low-cost lead-in fighter trainer and introduces a digital nav/attack system and other advanced avionics. First flying in December 1985, it retains the M.B.339B's engine and larger tip tanks, and is equipped to operate with stand-off weapons. Eighteen **M.B.339CB**s are operated by New Zealand. An anti-ship M.B.339C variant is known as the **M.B.339AM**, and is armed with Marte Mk 2 AShMs. In conjunction with Rolls-Royce and Lockeed, Aermacchi is offering the **T-bird II** as a contender for the US JPATS competition with uprated engine and improved avionics and structural modifications to extend service life.

The M.B.339C advanced fighter trainer can launch such stand-off weapons as Maverick ASMs, Marte 2 and AS34 Kormoran anti-ship missiles and laser-guided bombs.

Aermacchi M.B.339C

Specification: Aermacchi M.B.339C
Powerplant: one Piaggio-built Rolls-Royce Viper Mk 680-43 turbojet rated at 19.57 kN (4,400 lb st)
Dimensions: wing span over tip tanks 11.22 m (36 ft 9.75 in) ; length 11.24 m (36 ft 10.5 in); height 3.99 m (13 ft 1.25 in); wing area 19.30 m² (207.74 sq ft)
Weights: empty equipped 3310 kg (7,297 lb); normal take-off 4635 kg (10,218 lb); maximum take-off 6350 kg (13,999 lb)
Performance: maximum level speed 'clean' at sea level 487 kt (902 km/h; 560 mph); service ceiling 14630 m (48,000 ft); standard range 1,060 nm (1965 km; 1,221 miles); combat radius 270 nm (500 km; 311 miles) 'hi-lo-hi' profile with four Mk 82 bombs
Armament: maximum load of 1814 kg (4,000 lb) on six underwing hardpoints; ordnance includes 30-mm cannon pods, Miniguns, drop tanks, rockets, bombs, napalm, AIM-9P/L or Magic AAMs, AGM-65 AGMs, Marte Mk II AShMs and a four-camera reconnaissance pod

Aero L-39 Albatros

The L-39 was adopted as the standard advanced trainer of the former Warsaw Pact and was also widely exported. It can carry rocket pods and light bombs for weapons training.

The **Aero L-39 Albatros** two-seat basic and advanced trainer was developed with close co-operation from the USSR as a successor to the extensively built L-29 Delfin. It became the Warsaw Pact's standard jet trainer, emulating the success of its predecessor, and has been exported to many countries. Powered by a single Ivchenko AI-25 turbofan, the L-39 has an entirely conventional configuration with modular construction – there are only three major sub-assemblies (wing, fuselage and rear fuselage/tail). The type's maiden flight was made on 4 November 1968 and it entered Czech service in 1974.

Variants
More than 2,800 L-39s have been produced, including the **L-39C** basic and advanced trainer, **L-39V** target tug, **L-39ZO** weapons trainer with reinforced wings and four underwing weapon stations, and the **L-39ZA** for ground attack and reconnaissance, adding reinforced landing gear and an underfuselage gun pod to the L-39ZO.

Operators
The L-39 has been widely exported. Large numbers of L-39Cs remain in service in Russia and several other republics of the former USSR, and in Afghanistan, Algeria, Bulgaria, Congo, Cuba, Czechoslovakia, Egypt, Ethiopia, Hungary, Iraq, Libya, Nicaragua, Nigeria, North Korea, Romania, Syria and Vietnam.

L-59
Aero developed an improved **L-59** (originally **L-39MS**) variant, the first prototype making its maiden flight on 30 September 1986. The L-59 introduces a strengthened airframe, new avionics (including a HUD), revised flying controls and a new 22-kN (4,850-lb st) DV-2 turbofan. It is identifiable by its more sharply pointed nosecone and reshaped fin-tip. Deliveries of 48 **L-59E**s began in January 1993 to the Egyptian air force. These are equipped with US avionics. The **L-139** has been offered as a possible JPATS contender, powered by a Garrett TFE731 turbofan.

The L-39MS (now designated L-59) is fitted with a new DV-2 turbofan and has been exported to Egypt.

Aero L-39C Albatros

Specification: Aero L-59 (originally L-39MS)
Powerplant: one ZMDB Progress DV-2 turbofan rated at 21.57 kN (4,850 lb st)
Dimensions: wing span 9.54 m (31 ft 3.5 in) including tip tanks; length 12.20 m (40 ft 0.25 in); height 4.77 m (15 ft 7.75 in); wing area 18.80 m² (202.37 sq ft)
Weights: empty equipped 4150 kg (9,149 lb); normal take-off 5510 kg (12,147 lb) as a trainer; maximum take-off 5700 kg (12,566 lb) with external stores from a grass strip
Performance: maximum level speed 'clean' at 5000 m (16,405 ft) 473 kt (876 km/h; 544 mph); maximum rate of climb at sea level 1560 m (5,118 ft) per minute; service ceiling 11730 m (38,485 ft); ferry range at 9000 m (29,530 ft) 809 nm (1500 km; 932 miles) with drop tanks
Armament: underfuselage centreline point for podded 23-mm GSh-23 twin-barrelled cannon with 180 rounds, plus up to 1000 kg (2,204 lb) of weapons on four underwing pylons; including 500-kg (1,102 lb) bombs, 57-mm and 130-mm rocket pods, gun pods and two drop tanks

Agusta A 129 Mangusta

The A 129 Mangusta is the Italian army's standard light anti-tank and scout helicopter.

The **Agusta A 129 Mangusta** (Mongoose) was conceived to meet a 1972 Italian army requirement for a light anti-armour and scout helicopter. It was to be the first to incorporate a fully computerised and redundant integrated management system for a minimum crew workload. The A 129 features a conventional gunship layout of stepped tandem cockpits (pilot to rear and gunner in front), stub wings for weapons carriage, fixed, crashworthy undercarriage and slim fuselage for minimum visual signature. Composite materials account for 45 per cent of the fuselage weight.

Service

The first of five prototype A 129s, powered by two Piaggio-built Rolls-Royce Gem Mk 2-1004D turboshafts, made its maiden flight on 11 September 1983. An initial production batch of 15 from a requirement of 60 helicopters for the Italian army (ALE) was to have been delivered in late 1987. However, delays to allow fitting of a Saab/ESCO HeliTOW system with nose-mounted sight eventually put back initial delivery of five aircraft to October 1990. Two operational squadrons are planned within the ALE's 49° Gruppo at Casarsa. In 1993, the Mangusta made its first operational deployment in support of United Nations forces in Somalia.

Proposed variants

In 1986, an advanced A 129 variant was studied as a prospective European Light Attack Helicopter but was rejected after member countries pursued other programmes. Agusta has developed an export variant, which first flew in prototype form in October 1988. It features Allison/Garrett L800 LHTEC engines giving a power increase of over 20 per cent, uprated transmission and increased gross weight. A 10-12 passenger utility version has also been proposed as the **A 139**. Optional upgrades for export include a laser designator (for Hellfire missiles) and, for the scout role, a chin-mounted gun turret (which has already been qualified for use) and a mast-mounted sight.

A scout/attack Mangusta version has been proposed for escort and helicopter duties, fitted with a chin-mounted gun turret and armed with Stinger or Mistral air-to-air missiles.

Agusta A 129 Mangusta

Specification: Agusta A 129 Mangusta
Powerplant: two Rolls-Royce Gem 2 Mk 1004D turboshafts each rated at 615 kW (825 shp)
Dimensions: main rotor diameter 11.90 m (39 ft 0.5 in); wing span 3.20 m (10 ft 6 in); length overall, rotors turning 14.29 m (46 ft 10.5 in) and fuselage 12.28 m (40 ft 3.25 in); height overall 3.35 m (11 ft 0 in); main rotor disc area 111.22 m² (1,197.20 sq ft)
Weights: empty equipped 2529 kg (5,575 lb); maximum take-off 4100 kg (9,039 lb)
Performance: maximum level speed at sea level 140 kt (259 km/h; 161 mph); hovering ceiling 3750 m (12,300 ft) in ground effect and 3020 m (9,900 ft) out of ground effect; combat radius 54 nm (100 km; 62 miles) for a 90-minute patrol
Armament: up to 1200 kg (2,645 lb) of stores on four stub-wing pylons, including up to eight TOW-2A, HOT or six Hellfire ATGMs, Stinger, Mistral or AIM-9 AAMs, 52 70-mm (2.75-in) or 81-mm (3.18-in) SNIA-BPD rockets, and 7.62-, 12.7- or 20-mm (0.3-, 0.5- or 0.787-in) gun pods

AIDC Ching Kuo

Taiwan's Ching Kuo represents an ambitious attempt to develop an indigenous advanced combat aircraft, although much assistance has been provided by US companies.

Taiwan's ambitious programme to develop an advanced **Indigenous Defence Fighter (IDF)** to replace its fleet of F-5s and F-104s began in 1982 after a US arms embargo. However, no restrictions were placed on technical assistance, and US companies have collaborated closely with AIDC to develop the airframe (General Dynamics), radar (Westinghouse), engine (Garrett) and other systems. The resulting **Ching Kuo** fighter is of conventional all-metal construction and configuration, bearing a passing resemblance to an F-16/F-18 hybrid with wing/fuselage blending. Elliptical intakes are located below long leading-edge root extensions (LERXes) for good high-Alpha performance. The pressurised cockpit is fitted with a sidestick controller (like the F-16), a wide-angle HUD, three multi-function lookdown displays and an AN/APG-66-derived radar.

Indigenous weapons options

The first of three single-seat prototypes made its maiden flight on 28 May 1989, followed by the first two-seat prototype in July 1990. The first of 10 pre-production aircraft was rolled out on 9 March 1992 and introduced new enlarged engine intakes and a small ventral fin, following the loss of one of the prototypes in development flying. Deliveries to the air force began nearly one year earlier than scheduled with the public unveiling of the first squadron in February 1993. Weapons on display included GBU-12 500-lb laser-guided bombs, CBU-87 cluster munitions, AGM-65B TV Maverick ASMs and the indigenous Tien Chien I and Tien Chien II AAMs. The Tien Chien I closely resembles an AIM-9B or AIM-9D (though with wider span tailfins) while the semi-active radar-homing Tien Chien II is similar to the AIM-7 Sparrow in appearance. The Hsiung Feng II AShM may be derived from the Israeli Gabriel or AGM-84.

Reduced procurement

The original requirement called for 256 aircraft, including some trainers and some configured for anti-shipping duties. In March 1993 this procurement was reduced to 130 aircraft, to equip two, instead of the planned four, wings.

This view shows the elliptical air intakes, leading-edge flaps, wing LERXes and four indigenous Tien Chien I AAMs.

AIDC Ching Kuo

Specification: AIDC Ching Kuo
Powerplant: two ITEC (Garrett/AIDC) TFE1042-70 (F125) turbofans each rated at 26.80 kN (6,025 lb st) dry and 42.08 kN (9,460 lb st) with afterburning
Dimensions: wing span over wingtip missile rails (estimated) 8.53 m (28 ft 0 in); length including probe (estimated) 14.48 m (47 ft 6 in)
Weights: normal take-off 9072 kg (20,000 lb)
Performance: maximum level speed 'clean' at 10975 m (36,000 ft) more than 688 kt (1275 km/h; 792 mph); maximum rate of climb at sea level 15240 m (50,000 ft) per minute; service ceiling 16760 m (55,000 ft)
Armament: one internal 20-mm M61A1 cannon mounted beneath the port LERX; six hardpoints (two underfuselage, one under each wing and one wingtip missile rail) can carry ordnance including bombs, LGBs, cluster munitions, AIM-9P AAMs and indigenous IR-homing Tien Chien (Sky Sword) I, Sky Sword II AAMs (two in tandem recesses under the fuselage only) or three Hsiung Feng II (Male Bee II) anti-ship missiles

AMX International AMX

The AMX's potent all-weather precision attack capability has led to its description as the 'Pocket Tornado'.

AMX International AMX

Development of the **AMX** started in April 1978 when Aeritalia (now Alenia) and Aermacchi combined their resources to meet an Italian air force (AMI) requirement for an advanced multi-purpose strike/reconnaissance aircraft. The programme received extra impetus in 1980 when Brazil's EMBRAER joined the programme. A common specification called for good short-field performance, high subsonic operating speeds and advanced nav/attack systems for low-level day/night missions in poor visibility. This resulted in a conventional aircraft with a relatively compact airframe and moderately-swept high-mounted wings, and powered by a single Rolls-Royce Spey turbofan. An initial total of 266 aircraft was to be ordered, comprising 79 for Brazil, 187 for Italy and six single-seat prototypes. The first of these made the type's maiden flight in May 1984 in Italy and the first production aircraft entered service in 1989 with the Italian and Brazilian air forces.

Two-seat AMX-T

By 1989, programme totals had increased to 317 aircraft, with the addition of 51 two-seat **AMX-T**s to replace the AMI's ageing Fiat G91Ts. Retaining the same dimensions and full combat capabilities as the single-seat AMX, the trainer version replaces a fuel bay behind the original cockpit with a second ejection seat, with some reduction in range. Radar-equipped versions of the AMX-T are also under development for enhanced all-weather, ECR and maritime strike roles (armed with AM39 Exocet missiles).

Service

By mid-1993, Alenia and Aermacchi had delivered over 70 AMXs to equip five attack/reconnaissance squadrons with three wings, comprising 2°, 3° and 51° Stormi. A further three wings to equip it if full procurement is funded. By the same time, in Brazil, EMBRAER had delivered over 16 AMXs (locally designated **A-1**) which initially equipped 16° Grupo. Total Brazilian AMX procurement remains at 79, (including 14 AMX-T (**TA-1**) operational trainers) to equip four or five attack squadrons.

Brazil will acquire 14 TA-1 (AMX-T) operational trainers. Initial TA-1s have recently acquired USAF-style tailcodes.

Specification: AMX International AMX
Powerplant: one Fiat/Piaggio/Alfa Romeo Avio/CELMA-built Rolls-Royce Spey RB.168 Mk 807 turbofan rated at 49.06 kN (11,030 lb st)
Dimensions: wing span 8.87 m (29 ft 1.5 in) excluding wingtip missile rails and 10.00 m (32 ft 9.75 in) over wingtip AAMs; length 13.58 m (44 ft 6.5 in); height 4.58 m (15 ft 0.25 in); wing area 21.00 m² (226.05 sq ft)
Weights: operating empty 6700 kg (14,771 lb); normal take-off 9600 kg (21,164 lb); maximum take-off 13000 kg (28,660 lb)
Performance: maximum level speed 'clean' at 10975 m (36,000 ft) 493 kt (914 km/h; 568 mph); service ceiling 13000 m (42,650 ft); combat radius 480 nm (889 km; 553 miles) on a hi-lo-hi attack mission with a 907-kg (2,000-lb) warload
Armament: one internal M61A1 20-mm Vulcan cannon (Italy) and two 30-mm DEFA 554 cannon (Brazil); seven stores stations for up to 8,377 lb (3800 kg) of ordnance including bombs, ASMs, laser-guided munitions with targeting pods, self-defence AAMs (MAA-1 Piranha for Brazil)

Antonov An-12 'Cub'

The Shaanxi Y-8 is a reverse-engineered unlicensed An-12 copy, and serves with China in several specialised variants.

Designed as a high-wing, four-engined, rear-loading military freighter, the Ukrainian-built **Antonov An-12** has had considerable sales success in both civilian and military markets, and has been adapted to fulfil a variety of other roles. The prototype made its maiden flight during 1958. It is estimated that more than 900 were built at Kiev before production ceased during 1973, and more have been built under the designation **Shaanxi Y-8** in China, where the type is still produced and marketed. Developed from the passenger-carrying An-10, the **An-12BP** basic military freighter was designed from the start as a military transport and civilian freighter, with a rear-loading ramp, and a partly-pressurised cabin forward of the main portion of the freight compartment.The upswept rear fuselage consists of two split, inward-opening doors and a third upward-opening door aft to facilitate direct loading from trucks.

Specialised 'Cub' variants

Designated **'Cub'** by NATO, the An-12 has been produced in several versions and large numbers of military An-12s have been converted for specialist roles. **'Cub-A'** was a dedicated Elint platform with blade antennas on the forward fuselage. Still in front-line service, the Elint-gathering **'Cub-B'** has two prominent belly radomes and other blade antennas. **'Cub-C'** may be a dedicated ECM platform, with palletised electrical generators and control equipment, an array of antennas on its underside, various cooling scoops near the wing, and a bulged, ogival tailcone replacing the normal gun turret. **'Cub-D'** has a different equipment fit and is characterised by large external pods on the lower 'corners' of the forward fuselage and on each side of the base of the tailfin. Large numbers of An-12s have also been converted as one-off test and research platforms.

Customers

Large numbers of An-12s remain in service with the VTA. The An-12 has been exported to Algeria, Bangladesh, Egypt, Ethiopia, Guinea Republic, India, Iraq, Jordan, Madagascar, Poland, Syria, Yemen and Yugoslvia.

The 'Cub-D' is a second dedicated ECM platform. Unlike the 'Cub-C', it retains the conventional tail turret of other An-12s.

Antonov An-12 'Cub-A'

Specification: Antonov An-12BP 'Cub-A'
Powerplant: four ZMDB Progress (Ivchyenko) AI-20K turboprops each rated at 2983 kW (4,000 ehp)
Dimensions: wing span 38.00 m (124 ft 8 in); length 33.10 m (108 ft 7.25 in); height 10.53 m (34 ft 6.5 in); wing area 121.70 m² (1,310.01 sq ft)
Weights: empty 28000 kg (61,728 lb); normal take-off 55100 kg (121,473 lb); maximum take-off 61000 kg (134,480 lb); maximum payload 20000 kg (44,092 lb)
Performance: maximum level speed 'clean' at optimum altitude 419 kt (777 km/h; 482 mph); maximum cruising speed at opimum altitude 361 kt (670 km/h; 416 mph); maximum rate of climb at sea level 600 m (1,969 ft) per minute; service ceiling 10200 m (33,465 ft); take-off run 700 m (2,297 ft) at maximum take-off weight; landing run 500 m (1,640 ft) at normal landing weight; range 3,075 nm (5700 km; 3,542 miles) with maximum fuel or 1,942 nm (3600 km; 2,237 miles) with maximum payload
Armament: twin NR-23 23-mm cannon in tail turret

Antonov An-24/-26/-32

Libya has 10 An-26s for general transport duties. To ease loading from trucks, the rear-loading ramp of the 'Curl' can be slid forward along tracks to lie directly under the cabin.

The **Antonov An-24 'Coke'** tactical transport made its maiden flight on 20 December 1959, and was designed to meet an Aeroflot requirement for a turbine-engined replacement for piston-engined Il-14s and Il-12s. Its robustness, strength and performance appealed to military customers, and a total of 1,100 An-24s was built by the time production finished in 1978. The major production **An-24V** variant has seating for 28-40, side freight door and convertible cabin. Production continues in China as the **Xian Y-7**.

An-26 'Curl'

Although derived from the An-24, the **An-26 'Curl'** is a new design with a fully-pressurised cargo hold, uprated engines and a new rear-loading ramp to facilitate loading from trucks. All An-26s are fitted with an RU-19 turbojet in the rear of the starboard engine nacelle. As well as acting as an APU, this can be used as a take-off booster. A small number of An-26s have been converted as Elint/Sigint/EW platforms. These bear the NATO reporting name **'Curl-B'**, and have a profusion of swept blade antennas above and below the cabin. Production ended after about 1,000 had been built, mostly for military operators, but including 200 for Aeroflot and a handful more for civilian users in Afghanistan, Cuba, Laos, Mongolia, Romania, Syria and Yemen. The Chinese **Y-7H-500** remains in production.

'Cline'

The **An-32 'Cline'** replaced the An-26 in production, and is designed to offer improved take-off performance, ceiling and payload, especially under 'hot-and-high' conditions. The cabin can accommodate up to 50 passengers, 42 paratroops, or 24 stretcher patients and three attendants. All production aircraft are fitted with the 3812-ekW (5,112-ehp) AI-20D turboprops. These are mounted above the wing in very deep nacelles to give greater clearance for the increased-diameter propellers. As well as countries of the former USSR, the An-32 has already attracted a number of military customers, including Afghanistan, Bangladesh, Cuba, India, Mongolia and Peru.

India procured a total of 123 An-32 'Clines' to replace ageing C-119 and C-47 units. An-32s are named Sutlej after a river.

Antonov An-24RV 'Coke'

Specification: Antonov An-26B 'Curl-A'
Powerplant: two ZMDB Progress (Ivchyenko) AI-24VT turboprops each rated at 2103 kW (2,820 ehp) and one Soyuz (Tumanskii) RU-19A-300 turbojet rated at 7.85 kN (1,765 lb st)
Dimensions: wing span 29.20 m (95 ft 9.5 in); length 23.80 m (78 ft 1 in); height 8.58 m (28 ft 1.5 in); wing area 74.98 m² (807.10 sq ft)
Weights: empty 15400 kg (33,957 lb); normal take-off 23000 kg (50,705 lb); maximum take-off 24400 kg (53,790 lb); maximum payload 5500 kg (12,125 lb)
Performance: maximum level speed at 5000 m (16,405 ft) 291 kt (540 km/h; 336 mph); maximum level speed at sea level 275 kt (510 km/h; 317 mph); cruising speed at 6000 m (19,685 ft) 237 kt (440 km/h; 273 mph); range 1,376 nm (2550 km; 1,585 miles) with maximum fuel or 593 nm (1100 km; 683 miles) with maximum payload; take-off distance to 15 m (50 ft) 1240 m (4,068 ft) at maximum take-off weight; landing distance from 15 m (50 ft) 1740 m (5,709 ft)

Antonov An-124 'Condor'

The An-124 is similar in configuration to the C-5 Galaxy, but features a low-mounted tailplane. Still the world's largest production aircraft, its payload records remain intact.

Named after Pushkin's legendary giant, the **Ruslan** is in many respects comparable to the slightly smaller Lockheed C-5 Galaxy, which has a very similar configuration. Designed to meet Aeroflot and Soviet air force (VVS) Long-Range Transport Aviation (VTA) requirements for an An-22 replacement, the **Antonov An-124 'Condor'** has an upward-hinging 'visor-type' nose (with a folding nose ramp) and a large set of rear loading doors (with a three-part folding ramp) which allow simultaneous loading or unloading from both ends, or allow vehicles to be 'driven through'. The rear loading doors consist of the ramp, which can be locked in an intermediate position to allow direct loading from a truckbed, with an upward-hinging centre panel and downward-hinging clamshell doors behind. A handful of the 26 An-124s delivered by the end of 1992 are assigned directly to the VTA, wearing full military markings, but other Aeroflot aircraft are frequently employed on military tasks.

The An-124 remains the world's largest production aircraft (only the one-off six-engined **An-225** derivative is bigger), and has set a series of world records, most notably exceeding by 53 per cent the C-5's payload to 2000 m (6,560 ft). The vast, constant-section cargo hold has a titanium floor with rollgangs and retractable cargo tiedown points, and is lightly pressurised, with a fully-pressurised upper passenger deck for up to 88 people. For ease of loading the aircraft can be made to 'kneel' in a nose-down position by retracting the nosewheels and supporting the nose of the aircraft on retractable feet.

Fly-by-wire controls

The An-124 has fly-by-wire controls and a supercritical wing, and makes extensive use of composite materials for weight saving. The aircraft is capable of carrying virtually any load, including a complete SS-20 ICBM system, all Soviet main battle tanks, helicopters and other military equipment. In emergency, passengers can be carried in the lightly pressurised main hold, and in 1990 an An-124 carried 451 Bangladeshi refugees using foam rubber lining in the hold in lieu of seats.

The 150-tonne capacity An-124 is the VTA's primary heavy-lift strategic transport, and also serves with Aeroflot.

Antonov An-124 Ruslan

Specification: Antonov An-124 Ruslan 'Condor'
Powerplant: four ZMDB Progress (Lotarev) D-18T turbofans each rated at 229.47 kN (51,587 lb st)
Dimensions: wing span 73.30 m (240 ft 5.75 in); length 69.10 m (226 ft 8.5 in); height 20.78 m (68 ft 2.25 in); wing area 628.00 m² (6,759.96 sq ft)
Weights: operating empty 175000 kg (385,802 lb); maximum take-off 405000 kg (892,857 lb); internal fuel 230000 kg (507,055 lb); maximum payload 150000 kg (330,688 lb)
Performance: normal cruising speed at 10000 m (32,810 ft) between 432 and 459 kt (800 and 850 km/h; 497 and 528 mph); balanced take-off field length 3000 m (9,843 ft) at maximum take-off weight; landing run 800 m (2,625 ft) at maximum landing weight; range 9,140 nm (16500 km; 10,523 miles) with maximum fuel and 2,430 nm (4500 km; 2,796 miles) with maximum payload

Atlas Cheetah

The Cheetah upgrade represents an attempt to keep South Africa's surviving Mirage IIIs viable into the 21st century. The new wing and canard foreplanes greatly increase agility.

In July 1986 Atlas unveiled a considerably modified Dassault Mirage III, renamed **Cheetah**. Clearly benefiting from Israeli technology, the Cheetah bears some resemblance to the IAI Kfir. Conversion has been effected of the South African Air Force's (SAAF) later Mirage IIIs to similar standards, the IIICZs and IIIBZs being excluded. First to be upgraded were eight IIID2Z two-seaters, declared operational in August 1987 as the **Cheetah DZ**, and they have been followed by 14 **Cheetah EZ**s (Mirage IIIEZs), four RZs and three R2Zs, plus at least four DZs and five EZs produced from Mirages acquired from an undisclosed source. All ex-SAAF aircraft retain their original turbojet powerplants – Atar O9C-3s, apart from D2Zs with Atar O9K-50s – and the reconnaissance RZ/R2Zs also keep camera noses. All Cheetahs serve with No. 2 Squadron.

Cheetah modifications

Modifications to the airframe (including previously non-radar trainers which have an additional combat support role) include addition of Elta ranging radar in the nose, which is extended to accommodate additional avionics. The trainers have a Kfir TC2-type drooped nose, while Cheetah EZs feature a fuselage plug ahead of the windscreen. Installed internally are an Elbit HUD and INS including weapons computer, plus a locally-designed pilot's helmet sighting system. Provision is also made for a detachable refuelling probe mounted on the starboard air intake. The upgrade also adds two further hardpoints at the forward point of the wing/engine air intake trunk joint.

Both versions have Kfir-style canards and nose strakes, added as outward manifestations of an extensive structural strengthening programme. Wing leading-edge modifications are a 'dog-tooth' (a small notch) and replacement of the original slot by a small fence. These features give an invaluable increase in agility. The proposed Advanced Combat Wing has drooped leading edges and a kink outboard of the 'dog tooth', plus additional fuel capacity and wingtip missile rails. These changes have yet to be funded.

The two-seat Cheetah retains full combat capability, and may have a dedicated ground attack/interdiction role, where a backseater WSO may be useful.

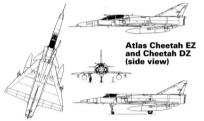

Atlas Cheetah EZ and Cheetah DZ (side view)

Specification: Atlas Cheetah EZ
Powerplant: one SNECMA Atar 9K-50 turbojet rated at 49.03 kN (11,023 lb st) dry and 70.82 kN (15,873 lb st) with afterburning
Dimensions: wing span 8.22 m (26 ft 11.6 in); canard foreplane span 3.73 m (12 ft 3 in); length including probe 15.65 m (51 ft 4.25 in); height 4.55 m (14 ft 11.25 in); wing area 34.80 m² (374.60 sq ft); canard foreplane area 1.66 m² (17.87 sq ft)
Performance: maximum level speed 'clean' at 12000 m (39,370 ft) 1,262 kt (2338 km/h; 1,453 mph); maximum cruising speed at 11000 m (36,090 ft) 516 kt (956 km/h; 594 mph); service ceiling 17000 m (55,775 ft)
Armament: two internal DEFA 30-mm cannon (single-seat aircraft only), plus up to 4000 kg (8,818 lb) of ordnance including V3B Kukri and V3C Darter IR AAMs, AS30 ASMs, laser-guided bombs, cluster munitions and rockets; typical weapons fit comprises eight 227-kg (500-lb) free-fall bombs, two V3B Kukris and two drop tanks

Beech C-12/RC-12

The antenna-festooned RC-12N Guardrail Common Sensors undertakes dual Comint/Elint missions. Guardrail is a communications intercept and direction finding system.

Beech C-12A Super King Air

The **Beech Super King Air 200** six/10-seat, twin-turboprop business aircraft was evolved from the King Air 100 as an enlarged, more powerful derivative. Introducing a T-tail, increased wing span, extra fuel and improved pressurisation, the Super King Air was to be adopted by all US armed services primarily as utility aircraft/light transports. These have the general designation **C-12/UC-12** and more than 300 are currently active. Navy and USMC UC-12s are engaged principally on routine communications and liaison tasks. Other military operators of the Super King Air 200 are Argentina, Bolivia, Canada, Chile, Colombia Ecuador, Egypt, Greece, Guatemala, Guyana, Hong Kong, Indonesia, the Irish Republic, Israel, Ivory Coast, Jamaica, Japan, Mexico, Morocco, Pakistan, Spain, Sri Lanka, Sweden, Thailand, Turkey and Venezuela. A radar-equipped maritime patrol version serves with Algeria, Peru and Uruguay.

RC-12 variants

Of special interest are several battlefield signals intelligence (Sigint) C-12 versions, usually with an extensive array of external antennas and comprehensive avionics. The first of these were **RU-21Js**, followed by the **RC-12D** with faired windows and a cabin packed with electronic recording equipment for Comint gathering. RC-12Ds are characterised by a wide array of dipole and blade antennas, and large wingtip ESM pods. Two similar versions are the **RC-12G** and **RC-12H** which differ in gross weight and equipment fit. The current major model is the **RC-12K** which combines the RC-12's Comint mission with the Elint mission previously undertaken by the Grumman RV-1D. Thirty-four are on order for the US Army, and deliveries began to Europe. These aircraft feature even larger numbers of aerials than their predecessors, with radar receivers having been added. The **RC-12N** is believed to be a similar version with some upgrades. Two US Navy RC-12s are configured as RANSAC (range surveillance aircraft) platforms with belly radomes. These comprise the **RC-12H** and **RC-12M** aircraft whose primary task is to check ranges for incursions by vessels prior to missile tests.

The US Marine Corps acquired 12 Beech UC-12Bs for use in liaison and base communications flight roles.

Specification: Beech RC-12K
Powerplant: two Pratt & Whitney Canada PT6A-41 turboprops each rated at 820 kW (1,100 shp)
Dimensions: wing span 17.63 m (57 ft 10 in) over ESM pods; length 13.34 m (43 ft 9 in); height 4.57 m (15 ft 0 in); wing area 28.15 m² (303.0 sq ft)
Weights: empty 3327 kg (7,334 lb); maximum take-off 7348 kg (16,200 lb); maximum payload more than 1043 kg (2,300 lb)
Performance: maximum level speed at 4265 m (14,000 ft) 260 kt (481 km/h; 299 mph); maximum cruising speed at 9145 m (30,000 ft) 236 kt (438 km/h; 272 mph); service ceiling 9420 m (30,900 ft); take-off distance to 15 m (50 ft) 869 m (2,850 ft); landing distance from 15 m (50 ft) 766 m (2,514 ft); range at maximum cruising speed 1,584 nm (2935 km; 1,824 miles)

Bell 205/212 UH-1D/N Iroquois

Bell's UH-1 Iroquois has achieved wide export sales. The Turkish armed forces operate Bell UH-1Hs on SAR (air force) and utility/VIP/transport (army) duties.

Bell UH-1N

Derived from the **Bell Model 204/UH-1 Iroquois**, the improved **Model 205** was first flown in August 1961. Retaining the existing T53 turboshaft, it introduced a larger-diameter main rotor, additional fuel capacity and a lengthened fuselage for a pilot and 12-14 troops, or six stretchers and a medical attendant, or 1814 kg (4,000 lb) of freight. Over 2,000 **UH-1D**s were built for the US Army, followed by the similar **UH-1H**, which introduced an uprated engine.

Specialised UH-1H variants include the **EH-1H** ECM jammer, **JUH-1H** radar testbed and **UH-IV** medevac/rescue helicopter. Other versions are the Canadian **CUH-1H** operational trainer and the **HH-1H** rescue helicopter for the USAF. Some 3,573 UH-1Hs were built, and it is planned to retain large numbers for service into the 21st century. The type has been exported to nearly 50 countries.

US Army Hueys have been upgraded with new avionics and equipment; new composite main rotor blades are to be introduced, as well as an improved cockpit. Many export customers are also looking to upgrade their ageing Hueys. Three different programmes have been offered, primarily centred around engine and dynamic system upgrades to improve payload/range capability. The **Bell UH-1HP Huey II** has an uprated T53 giving a 300-kW (400-shp) power increase; **Huey 800** replaces the T53 with an LHTEC T800-800 engine and UNC Helicopter's **UH-1/T700 Ultra Huey** uses the same airframe/dynamic system improvements as the Huey II, but with the Sikorsky UH-60's 1400-kW (1,900-shp) GE T700 engine.

Twin-turbine Model 212/UH-1N

The **Model 212** is a twin-turbine UH-1H, fitted with two PT6T turboshafts mounted side-by-side to drive a single shaft. The USAF's 79 **UH-1N**s saw service in support of Special Operations COIN activities. The principal USN and USMC variant is the **HH-1N**, and is used for utility/SAR (USN) and combined transport/light attack (USMC) duties. Agusta produced the type under licence as the **AB 212**, and has developed the specialised maritime **AB 212ASW**. This is equipped with a Bendix sonar or search radar for the ASV role, and can carry torpedoes and AShMs.

Argentina operates Bell 212s on SAR/transport duties.

Specification: Bell Helicopter Textron Model 205/UH-1H Iroquois
Powerplant: one Textron Lycoming T53-L-13 turboshaft rated at 1044 kW (1,400 shp)
Dimensions: main rotor diameter 14.63 m (48 ft 0 in); length overall, rotors turning 17.62 m (57 ft 9.625 in) and fuselage 12.77 m (41 ft 10.25 in); height overall 4.41 m (14 ft 5.5 in) with tail rotor turning; main rotor disc area 168.11 m² (1,809.56 sq ft)
Weights: empty equipped 2363 kg (5,210 lb); basic operating 2520 kg (5,557 lb) in troop-carrying role; normal take-off 4100 kg (9,039 lb); maximum take-off 4309 kg (9,500 lb); maximum payload 3,880 lb (1759 kg)
Performance: maximum level, maximum cruising and economical cruising speed at 1735 m (5,700 ft) 110 kt (204 km/h; 127 mph); maximum rate of climb at sea level 488 m (1,600 ft) per minute; service ceiling 3840 m (12,600 ft); hovering ceiling 4145 m (13,600 ft) in ground effect and 1220 m (4,000 ft) out of ground effect; range 276 nm (511 km; 318 miles)

Bell 206/OH-58 Combat Scout

Utility/armed scout helicopter

Although its primary role is to act as an airborne scout and laser designator for other aircraft, the OH-58D can also carry out autonomous attacks with Hellfire ATGMs and gun pods.

Bell OH-58D Kiowa Warrior

In March 1968 Bell's **Model 206A JetRanger** five-seat light helicopter was ordered into production for the US Army as the **OH-58A Kiowa** for the light observation role. Deliveries to the US Army began on 23 May 1969, and over five years a total of 2,200 was procured. Export customers comprised Australia (56 licence-built **Kalkadoon**s), Austria (12 **OH-58B**s) and Canada (74 **COH-58A**s).

Improved and armed OH-58 variants

From 1978, 585 OH-58As were converted to improved **OH-58C** standard with a flat glass canopy, an uprated engine with infra-red suppression and improved avionics. In 1981, the **Bell Model 406** proposal was developed as a scout helicopter. Designated **OH-58D**, it introduced a mast-mounted sight, four-bladed rotor, specialised avionics and a new cockpit display system. Initial plans for upgrading 592 US Army OH-58As to 'D' standard were progressively reduced to the current total of 363. Fifteen armed **Prime Chance OH-58D**s were modified in 1987 for operations against Iranian fast-patrol boats in the Persian Gulf. The armament options were retained for an armed OH-58D, designated **Kiowa Warrior**, to which standard 243 planned OH-58Ds are to be modified with an integrated weapons pylon, uprated engine and transmission, increased gross weight, RWR, IR jammer, laser warning receiver, integrated avionics and a lightened structure. New-build Kiowa Warriors were delivered to the US Army in May 1991.

Combat Scout

Eighty-one Kiowa Warriors are to be modified further as **Multi-Purpose Light Helicopter**s (MPLH) with squatting landing gear and quick-fold rotor blades, horizontal stabiliser and tilting fin for transportation in C-130s and rapid redeployment. The **Model 406 CS Combat Scout** is a lighter and simplified export derivative of the OH-58D, retaining the main rotor, tail rotor and transmission and a similar powerplant. Fifteen TOW-capable OH-58s were delivered from June 1990 onwards to Saudi Arabia as **MH-58D**s.

The mast-mounted sight provides all-weather targeting information and is boresighted to a laser rangefinder.

Specification: Bell Helicopter Textron Model 406/OH-58D Kiowa Prime Chance
Powerplant: one Allison T73-AD-700 turboshaft rated at 485 kW (650 shp)
Dimensions: main rotor diameter 10.67 m (35 ft 0 in); length overall, rotors turning 12.85 m (42 ft 2 in) and fuselage 10.48 m (34 ft 4.75 in); height overall 3.93 m (12 ft 10.625 in); main rotor disc area 89.37 m² (962.00 sq ft)
Weights: empty 1381 kg (3,045 lb); maximum take-off 2041 kg (4,500 lb)
Performance: maximum level speed 'clean' at 1220 m (4,000 ft) 128 kt (237 km/h; 147 mph); hovering ceiling more than 3660 m (12,000 ft) in ground effect and 3415 m (11,200 ft) out of ground effect; range 250 nm (463 km; 288 miles)
Armament: 12.7-mm (0.50-in) machine-guns, seven-tube 70-mm rocket pods, plus provision for Stinger IR AAMs and Hellfire anti-armour missiles; Combat Scout armament options include two GIAT 20-mm cannon pods, four TOW 2 or Hellfire anti-armour missiles, or Stinger missiles, 70-mm rockets and 7.62-mm or 12.7-mm guns

Bell 209/AH-1 HueyCobra

The AH-1W is the USMC's standard attack helicopter, and currently represents the ultimate 'Cobra variant in service. An upgrade programme will provide a night targeting system.

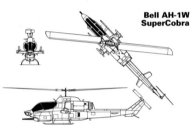

**Bell AH-1W
SuperCobra**

Produced in 1965 as an 'interim' armed helicopter escort, the **Bell Model 209 HueyCobra** was derived from the UH-1 assault helicopter, retaining its powerplant, transmission and rotor, but introducing a new, slimmer fuselage with now-standard 'gunship' configuration of stepped tandem cockpits. Bell produced some 1,100 **AH-1G**s for US Army service in Vietnam. Some were later upgraded to TOW-capable **AH-1Q** standard. A planned improvement programme to update existing airframes to '**AH-1S(MC)**' standard was never fully achieved, leaving airframes in different modification states. In 1989 a redesignation exercise renamed the MC '**AH-1F**', and less advanced versions as **AH-1P**, **AH-1E** and **AH-1S**. The latter is the current production configuration and thus applies to most export machines. Most are distinguishable by the 'flat-plate' canopy, TOW sighting system nose turret, prominent exhaust suppressor and two-bladed composite rotor with tapered tips. Primary armament is four TOWs and a chin-mounted 20-mm cannon. The largest user of the AH-1F/S remains the US Army, with some 800 left in service. The 'F' is being built under licence in Japan and has been exported to Israel, Jordan, Pakistan, South Korea, and Thailand and Turkey. A handful of AH-1S aircraft remain in use, as well as some AH-1Ps and AH-1Es.

Twin-engined variants
Developed for the USMC, the **AH-1J** was essentially a twin-engined cannon-armed AH-1G. There followed the improved Iranian **AH-1J** and the stretched **AH-1T/W** models with a lengthened tail boom and forward fuselage plug for the larger main rotor. The AH-1W has GE T700 engines and 'cheek bulges' housing TOW 'black boxes'. The USMC is the primary user and will eventually receive 230 (including the rebuilt Tango model). Foreign operators are Thailand and Turkey. The **AH-1(4B) W Viper** is an intended upgrade standard with advanced cockpit displays, uprated T700s and two-bladed rotor replaced with a more advanced four-bladed design. The AH-1W also forms the basis of the **CobraVenom** variant offered by Bell and GEC Avionics to meet the UK requirement for a new battlefield helicopter.

The AH-1S model is distinguished by its 'flat-plate' canopy.

Specification: Bell Helicopter Textron Model 209 (AH-1W SuperCobra)
Powerplant: two 1212-kW (1,625-shp) General Electric T700-GE-401 turboshafts; transmission limited to 1515 kW (2,032 shp) for take-off and 1286 kW (1,725 shp) for continuous running
Dimensions: main rotor diameter 14.63 m (48 ft 0 in); wing span 3.23 m (10 ft 7 in); length overall, rotors turning 17.68 m (58 ft 0 in) and fuselage 13.87 m (45 ft 6 in); height overall 4.32 m (14 ft 2 in); main rotor disc area 168.11 m² (1,809.56 sq ft)
Weights: empty 4627 kg (10,200 lb); maximum take-off 6691 kg (14,750 lb)
Performance: maximum level speed 'clean' at sea level 152 kt (282 km/h; 175 mph); range 343 nm (635 km; 395 miles) with standard fuel
Armament: one chin-mounted M-197 three-barrelled 20-mm cannon; maximum ordnance 1119 kg (2,466 lb) including eight TOW or Hellfire ATGMs, seven- or 19-shot 70-mm (2.75-in) rocket pods, 127-mm (5-in) Zuni rockets, cluster munitions, napalm, AIM-9 and Stinger IR AAMs and drop tanks; qualified for AGM-65 Maverick AGMs

Bell/Boeing V-22 Osprey

The No. 2 V-22 prototype approaches a refuelling drogue towed by a KC-130 Hercules tanker. The engine nacelles are tilted almost vertically to improve low-speed control.

Bell Helicopter Textron and Boeing Vertol joined forces in the early 1980s to develop a larger derivative of the XV-15 tilit-rotor demonstrator for the Joint Services Advanced Vertical Lift Aircraft (formerly **JVX**) programme. Combining the vertical lift capabilities of a helicopter with the fast-cruise (275 kt) forward flight efficiencies of a fixed-wing turboprop aircraft, the resulting **V-22 Osprey** was awarded full-scale development in 1985. It is powered by two Allison T406 turboshaft engines driving three-bladed prop-rotors through interconnected drive shafts. The wingtip-mounted engine nacelles can be swivelled through 97.5°.

Proposed variants

Initial requirements called for 913 Ospreys, comprising 552 **MV-22A** USMC assault versions carrying 24 fully-armed troops; 231 similar variants for the US Army; 80 USAF **CV-22A**s for long-range transport of special forces personnel; and 50 USN **HV-22A**s for combat SAR, special warfare and fleet logistic support with a 9072-kg (20,000-lb) payload. The USN also foresaw a need for up to 300 **SV-22A** ASW versions. For shipboard stowage, the V-22 mainplanes pivot centrally to rotate along the fuselage top, the prop-rotor blades also folding in parallel.

Development

Flight-testing started on 19 March 1989 and the Osprey successfully demonstrated airborne transition from heli-copter to wing-borne flight in September 1989. Funding for further development was cut in 1990 and all except the USMC requirement was deleted, and in mid-1992 this was reduced to a baseline figure of 300 Ospreys. Even after consideration of competing helicopters through the USMC's Medium Lift Replacement (MLR) programme, the Bell/Boeing team was awarded a US Navy contract for another four V-22s, after modifying two of the existing prototypes to similar lighter and cheaper 'production-repre-sentative' standards. These six Ospreys are to be evaluated until 1998 against several medium-lift helicopter projects, including the Boeing CH-47D/F, EH.101 and Sikorsky S-92.

This prototype V-22 was painted in USMC colours. The service remains committed to the controversial programme.

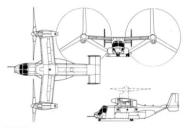

Bell/Boeing V-22 Osprey

Specification: Bell/Boeing V-22 Osprey
Powerplant: two Allison T406-AD-400 turboshafts each rated at 4586 kW (6,150 shp) take-off and 4392 kW (5,890 shp) continuous running
Dimensions: rotor diameter, each 11.58 m (38 ft 0 in); wing span 15.52 m (50 ft 11 in) including nacelles; length, fuselage excluding probe 17.47 m (57 ft 4 in); height over fins 5.38 m (17 ft 7.8 in); wing area 35.59 m² (382.00 sq ft); total rotor disc area 210.72 m² (2,268.23 sq ft)
Weights: empty equipped 14463 kg (31,886 lb); maximum take-off 27442 kg (60,500 lb) for STO; maximum internal payload 9072 kg (20,000 lb); maximum external payload 6804 kg (15,000 lb)
Performance: maximum cruising speed at sea level 100 kt (185 km/h; 115 mph) in helicopter mode and 275 kt (509 km/h; 316 mph) in aeroplane mode; service ceiling 7925 m (26,000 ft); tactical range 1,200 nm (2224 km; 1,382 miles) after VTO at 21146 kg (44,619 lb) with 5443-kg (12,000-lb) payload

Boeing B-52G/H Stratofortress

A B-52H of the 23rd Bomb Squadron, 5th Bomb Wing. Known unofficially as the 'Cadillac', about ninety turbofan-powered H models serve with Air Combat Command.

Backbone of SAC, the **Boeing B-52** first entered service in June 1955. Only two versions remain in service with its successor, ACC. A total of 193 J57-powered **B-52G**s was built with integral wing tanks and short fin, and armed with four tail-mounted 0.50-in machine-guns. As power projection is their principal role, Desert Storm missions were all assigned to the G model operating as free-fall bombers. Some nuclear strike-configured aircraft were converted to carry 12 AGM-86B cruise missiles underwing. However, conventional AGM-86Cs are now in use. Non-cruise configured B-52Gs were assigned a maritime role with Harpoon ASMs. The B-52G is well-protected by numerous ECM systems, and two undernose blisters house LLLTV and FLIR sensors, which are used with terrain-avoidance radar to provide low-level penetration capability in bad weather or at night. The reduction in post-Cold War nuclear threat has released B-52Hs to assume many of the conventional tasks previously undertaken by the B-52G. Consequently, the G model will be out of service by the mid-1990s.

Turbofan-powered model

The **B-52H** was intended to serve as a Skybolt missile carrier. It was then assigned a low-altitude all-weather penetration role, resulting in major structural strengthening. Although based on the B-52G, it replaced turbojets with TF33 turbofans, giving a significant range increase, and introduced a six-barreled 20-mm Vulcan cannon. The first of 102 production B-52Hs was flown on 6 March 1961. Since then update programmes have added improved avionics, ECM protection and an Electro-optical Viewing System (EVS). B-52Hs were configured to carry 20 AGM-86B cruise missiles (12 under the wing pylons and eight on an internal rotary launcher), and were employed mainly in the stand-off nuclear missile launch role until 1991, when the force began to adopt a wider brief including conventional bombing tasks. It has also added the stealthy AGM-129 Advanced Cruise Missile to its inventory. Approximately 90 B-52Hs remain in the inventory and, unlike the G models, no plans have been announced for any retirements.

The B-52G was heavily committed to Desert Storm missions, delivering one third of ordnance dropped by USAF aircraft.

Boeing B-52H Stratofortress

Specification: Boeing B-52H Stratofortress
Powerplant: eight Pratt & Whitney TF33-P-3 turbofans each rated at 75.62 kN (17,000 lb st)
Dimensions: wing span 56.39 m (185 ft 0 in); length 49.05 m (160 ft 10.9 in); height 12.40 m (40 ft 8 in); wing area 371.60 m² (4,000.00 sq ft)
Weights: maximum take-off 229088 kg (505,000 lb)
Performance: cruising speed at high altitude 442 kt (819 km/h; 509 mph); penetration speed at low altitude between 352 and 365 kt (652 and 676 km/h; 405 and 420 mph); service ceiling 16765 m (55,000 ft); take-off run 2896 m (9,500 ft) at maximum take-off weight; range more than 6,865 nm (16093 km; 10,000 miles)
Armament: one 20-mm Vulcan six-barrelled cannon in tail turret housing, plus up to 22680 kg (50,000 lb) of ordnance including AGM-86C conventional-warhead cruise missiles, B61 or B83 nuclear weapons, AGM-142 Have Nap (Rafael Popeye) precision-guided attack missiles and up to 51 340-kg (750-lb) or 454-kg (1,000-lb) Mk 83 conventional bombs

Boeing KC-135 Stratotanker

USA
Inflight-refuelling tanker

A KC-135R of the 92nd Bomb Wing refuels the first YF-22 during flight trials from Edwards. Tanker support enabled each test sortie to be as long and productive as possible.

Following successful trials of the Boeing 367-80 transport prototype with a Boeing-designed flying boom under the rear fuselage, an order was received for **KC-135A** tankers in September 1955. The first flew in August 1956 and USAF service entry followed in 1957. A total of 732 KC-135s was built, 582 with a short fin, after which a taller fin was introduced (and retrofitted) for improved take-off stability. Internally, the KC-135A features integral wing and lower fuselage fuel tanks. A side-loading cargo door is fitted for cargo carriage. Alternatively, seating can be provided for 80 troops.

Fifty-six KC-135As were built as **KC-135Q**s for SR-71 support with additional nav/comms equipment and carrying high-flashpoint JP-7 fuel in addition to the regular JP-4/5. Other early tanker variants were the **C-135F** (12 KC-135As supplied to France), and the **KC-135D**, four of which were converted from RC-135A survey aircraft. By early 1993 the original J57-engined variant was serving with 12 ACC units.

Turbofan engines

Two major turbofan re-engining programmes have been undertaken. The first involved conversion of 163 ANG and AFRES KC-135s, four KC-135Ds and 21 special-mission aircraft to **KC-135E** standard with TF33 turbofans and widerspan tailplanes (both from surplus 707 airliners). The thrust reverser-equipped TF33 allows greater safety margins, use of shorter runways, reduced nosie pollution and greater fuel offloads on similar mission profiles. The Boeing-developed **KC-135R** first flew in August 1982 and is now the mainstay of the US Air Force's tanker fleet. Over 300 CFM56-powered conversions have been funded so far, with the first entering service in July 1984. The surviving 11 C-135Fs were also similarly upgraded to become **C-135FR**s. Related variants include the **KC-135R(RT)**, applied to a small number of mostly ex-special mission or trials aircraft with a refuelling receptacle. Survivors of the 56 KC-135Q aircraft are being fitted with refuelling receptacles as they undergo the re-engining conversion to emerge as **KC-135T**s, with a primary role of supporting F-117 attack aircraft and other covert programmes.

The J57-engined KC-135A was the main production version. Many have been re-engined with TF33 or CFM56 turbofans.

Boeing KC-135R Stratotanker

Specification: Boeing KC-135R Stratotanker
Powerplant: four CFM International F108-CF-100 turbofans each rated at 97.86 kN (22,000 lb st)
Dimensions: wing span 39.88 m (130 ft 10 in); length 41.53 m (136 ft 3 in); height 12.70 m (41 ft 8 in); wing area 226.03 m² (2,433.00 sq ft)
Weights: operating empty 48220 kg (106,306 lb); maximum take-off 146284 kg (322,500 lb); internal fuel 92210 kg (203,288 lb); maximum payload 37650 kg (83,000 lb)
Performance: maximum level speed at high altitude 530 kt (982 km/h; 610 mph); cruising speed at 10670 m (35,000 ft) 462 kt (856 km/h; 532 mph); maximum rate of climb at sea level 393 m (1,290 ft) per minute; service ceiling 13715 m (45,000 ft); typical take-off run 3261 m (10,700 ft) at maximum take-off weight; operational radius 2,500 nm (4633 km; 2,879 miles) to offload 150 per cent more fuel than the KC-135A – KC-135A operational radius 3,000 nm (5560 km; 3,455 miles) to offload 10886 kg (24,000 lb) of fuel or 1854 km (1,151 miles: 1,000 nm) to offload 54432 kg (120,000 lb) of fuel

21

Boeing C-135 special variants

A sizeable fleet of C-135 airframes is used for test purposes. This NKC-135A refuels a NASA SR-71B over Edwards AFB.

A total of 820 **Boeing C-135**s was built. Although the KC-135 tankers were the principal variants, 45 were built as **C-135A/B** transports, with no tanking equipment. Several of these remain in limited service on specialised transport and test duties, including turbofan re-engined **C-135E**s. Three **WC-135B** weather reconnaissance aircraft have reverted to transport status, becoming **C-135C**s. Surplus 707 airliners have also served with the USAF as VIP transports; these are designated **VC-137A** (re-engined as the **VC-137B**), **C-137C** (including two Presidential transports) and **EC-137D**. Four more were modified to **EC-18B** standard for the ARIA (Advanced Range Instrumentation Aircraft) role. Two are equipped as **EC-18D** Cruise Missile Mission Control Aircraft.

EC-135 airborne command post and test aircraft

The **EC-135** designation applies to dedicated Airborne Command Post mission aircraft equipped with comprehensive communications equipment. Backbone of the fleet is the **EC-135C**, serving with the 55th Wing. The **EC-135A**, **G**, **J** and **L** versions have been retired. Further aircraft include **EC-135K**s used to provide navigation support for fighter deployments, and **EC-135E** range support/test aircraft. Under various designations, including **NC-135A**, **NKC-135A** and **NKC-135E**, grossly modified USAF C-135s are operated on development and trials work. Two jammer-equipped NKC-135As serve with the USN on EW support duties.

RC-135 reconnaissance variant

The **RC-135** designation applies to reconnaissance variants. All have much on-board electronic recording and analysing equipment. Three Sigint-gathering variants have slab-sided antenna cheek fairings. These comprise the **RC-135U**, the **RC-135V 'Rivet Joint'** with 'thimble' nose and the essentially similar **RC-135W** with large plate aerials. The related **TC-135W** is a crew trainer. The telemetry intelligence role is undertaken by **RC-135S 'Cobra Ball'** aircraft, optimised for photography of foreign missile tests. A single **TC-135S** provides aircrew training for the Telint fleet.

The USAF's RC-135V/W 'Rivet Joint' fleet undertakes strategic reconnaissance missions. The slab-sided cheek fairings contain electronic receivers.

Boeing NKC-135A

Specification: Boeing EC-135C
Powerplant: four Pratt & Whitney TF33-P-9 turbofans each rated at 80.07 kN (18,000 lb st)
Dimensions: wing span 39.88 m (130 ft 10 in); length 41.53 m (136 ft 3 in); height 12.70 m (41 ft 8 in); wheel base 13.92 m (45 ft 8 in); wing area 226.03 m² (2,433.00 sq ft)
Weights: basic empty 46403 kg (102,300 lb); maximum take-off 135626 kg (299,000 lb)
Performance: maximum level speed at 7620 m (25,000 ft) 535 kt (991 km/h; 616 mph); cruising speed at 10670 m (35,000 ft) 486 kt (901 km/h; 560 mph);maximum rate of climb at sea level 610 m (2,000 ft) per minute; service ceiling 12375 m (40,600 ft); ferry range 4,910 nm (9099 km; 5,654 miles); operational radius 2,325 nm (4308 km; 2,677 miles)

Boeing E-3 Sentry

Perhaps the most advanced Sentries are the RAF's E-3D Sentry AEW.Mk 1s, powered by CFM56 turbofans and equipped with passive sensors in the wingtip pods.

The **Boeing E-3 Sentry** is the West's principal AWACS (airborne warning and control system) platform. The **EC-137D** prototype first flew on 5 February 1972, followed by the first **E-3A** on 31 October 1975. Using the airframe of a 707-320B airliner and a massive payload of radar and electronic sensors, the E-3 is a flying headquarters for C^3I, employed near a combat zone to monitor aircraft and missiles and to direct friendly warplanes. Heart of the system is an AN/APY-2 Overland Downlook Radar (ODR) which is mounted (along with other sensors and instrumentation) in a rotodome above the rear fuselage. The radar is capable of tracking up to 600 low-flying aircraft. Since entering service, AWACS aircraft have been involved in combat operations in Grenada (1983), Lebanon (1983), Panama (1989) and Iraq (1991) and continuing operations over Bosnia (1993/94).

Upgraded E-3B/Cs

Twenty-two E-3As and two EC-137Ds, collectively termed 'core' aircraft when they were standardised in the late 1970s, were upgraded to **E-3B** standard with faster computing ability, ECM-resistant communications and additional radios and display consoles. The first E-3B was redelivered in July 1984. In 1984, 10 E-3As were modified to **E-3C** standard with slightly larger crew capacity, most E-3B equipment and Have Quick communications equipment. All but the first 25 E-3 airframes have inboard underwing hardpoints. E-3A 'standard' versions have been delivered to Saudia Arabia (five) and NATO (18).

Re-engined export E-3s

Sentries have been exported to both the UK and France, which operate seven **E-3D Sentry AEW. Mk 1** and four **E-3F SDA** (Système de Détection Aéroportée) aircraft respectively. Both E-3 models entered service in 1991 and have 106.8-kN (24,000-lb st) CFM56 turbofans and SOGERMA IFR probes in addition to the flight-refuelling receptacle. RAF aircraft also have wingtip-mounted Loral Yellow Gate ESM pods. CFM56 engines power all five Saudi E-3As and eight **KE-3A** dedicated tanker aircraft.

The USAF received 34 E-3s, comprising 25 E-3As (ugraded to E-3B standard with improved radar) and nine E-3Cs.

Boeing E-3A Sentry

Specification: Boeing E-3C Sentry
Powerplant: four Pratt & Whitney TF33-P-100/100A turbofans each rated at 93.41 kN (21,000 lb st)
Dimensions: wing span 44.42 m (145 ft 9 in); length 46.61 m (152 ft 11 in); height 12.73 m (41 ft 9 in); wing area 283.35 m² (3,050.00 sq ft)
Weights: operating empty 77996 kg (171,950 lb); maximum take-off 147420 kg (325,000 lb); internal fuel 90800 litres (23,987 US gal)
Performance: maximum level speed at high altitude 460 kt (853 km/h; 530 mph); operating ceiling 8840 m (29,000 ft); operational radius 870 nm (1612 km; 1,0002 miles) for a 6-hour patrol without flight refuelling; endurance more than 11 hours without flight refuelling

Boeing/Grumman E-8 J-STARS

Battlefield control aircraft

A converted 707-320C airliner, the E-8A battlefield surveillance platform uses a Norden multi-mode radar as its primary sensor. The darker grey aircraft (background) is the first conversion.

Making a 'star' appearance in Operation Desert Storm long before it was considered operational, the **Boeing/Grumman E-8** represents a major advance in battlefield control, introducing the kind of capability for monitoring and controlling the land battle that the E-3 provides for the air battle. Like the E-3, the E-8 is based on the Boeing 707-320 airliner airframe, and no new-build aircraft are envisaged.

Mission equipment

Two **E-8A** prototypes were converted, with the first flying in operational configuration in December 1988. They introduced a Norden multi-mode SLAR (housed in a forward fuselage ventral canoe fairing) and a cabin configured with operator consoles. The radar can operate in synthetic aperture mode, which gives a high resolution radar picture out to 257 km (160 miles) from the orbiting aircraft, while two pulse-Doppler modes give moving target information. Wide area search/moving target indicator mode monitors a large sector of land, while sector search mode is used on much smaller areas to follow individual vehicles. A datalink is used to transmit intelligence gathered in near real-time to mobile ground consoles, similar to those on the E-8. Using the various modes, the J-STARS (Joint Surveillance Target Attack Radar System) can be used for general surveillance and battlefield monitoring to provide the 'big picture' to commanders, stand-off radar reconnaissance or individual targeting functions for attacking vehicles and convoys.

In January 1991 both E-8As deployed to Riyadh to fly combat missions. Forty-nine war missions were flown, for a total of 535 hours, a sizeable portion of which was spent on the search for Iraqi 'Scud' missiles. Desert Storm provided ground and air commanders with a wealth of material, and fully validated the concept.

In service, the system was to have been carried on the new-build **E-8B** aircraft with F108 turbofans but, despite one **YE-8B** being procured (later sold), the carrier platform will now be the **E-8C** based on converted 707 airliner airframes. A total of 20 is required, the first of which was funded in 1990. Initial deliveries are expected in 1995.

During Desert Storm the two E-8As operated from Riyadh. Their contribution to the coalition success was enormous.

Boeing/Grumman E-8A J-STARS

Specification: Boeing/Grumman E-8A
Powerplant: four Pratt & Whitney JT3D-7 turbofans each rated at 84.52 kN (19,000 lb)
Dimensions: wing span 44.42 m (145 ft 9 in); tailplane span 13.95 m (45 ft 9 in); length 46.61 m (152 ft 11 in); height 12.93 m (42 ft 5 in); wheel base 17.98 m (59 ft 0 in); wing area 283.35 m² (3,050.00 sq ft)
Weights: maximum take-off 151315 kg (333,600 lb)
Performance: maximum cruising speed at 7620 m (25,000 ft) 525 kt (973 km/h; 605 mph); economical cruising speed at 10670 m (35,000 ft) 464 kt (860 km/h; 534 mph); maximum rate of climb at sea level 1219 m (4,000 ft) per minute; service ceiling 11890 m (39,000 ft); range with maximum fuel 5,000 nm (9266 km; 5,758 miles)

Boeing/Sikorsky RAH-66 Comanche

This Comanche mock-up shows the futuristic shape and faceted structure. Hellfire anti-tank missiles can be carried internally, and on pylons under detachable stub wings.

The US Army issued its LHX (Light Helicopter Experimental) requirement in 1982, initially calling for 5,000 helicopters to replace UH-1, AH-1, OH-6 and OH-58 scout/attack/assault aircraft. This has since been scaled down to 1,292 for the scout/attack role only. Boeing/Sikorsky's 'First Team' was awarded the contract (over Bell/McDonnell Douglas 'Super Team') for three dem/val aircraft on 5 April 1991.

Boeing/Sikorsky RAH-66 Comanche

Configuration

The **RAH-66 Comanche** has a five-bladed all-composites bearingless main rotor and an eight-bladed fan-in-fin shrouded tail rotor. Its largely composite airframe is designed for low observability, employing some degree of faceting and sunken notch intakes for the two LHTEC T800 turboshafts. The undercarriage is retractable, and all weapons are housed internally, with missiles carried in bays on the fuselage sides, directly attached to the bay doors which act as pylons when they are open. A chin turret houses a 20-mm cannon, and in the extreme nose is a sensor turret for a FLIR and a laser designator.

Advanced avionics

The Army has specified maximum avionics commonality with the USAF's F-22 ATF and the pilot (in front) and WSO each have two flat screen MFDs for presentation of tactical situation, moving map and FLIR/TV information. The pilot also has a wide field-of-view helmet-mounted display system, allied to Martin Marietta electro-optical night navigation and targeting systems. Flight control is by a triplex fly-by-wire system, with sidestick cyclic-pitch controls. The RAH-66 also features a wide array of defensive equipment, including laser-, IR- and radar-warning receivers.

The US Army's RAH-66 specified empty weight of 3402 kg (7,500 lb) has grown by nearly one quarter as a result of additional equipment (including a miniaturised version of the Longbow MMW radar specified for the AH-64D Apache). The First Team's aircraft is expected to fly in August 1995, with expected IOC in late 1998.

Under the 'First Team' partnership, Boeing is responsible for the main rotor blades and tail section, while Sikorsky builds the forward fuselage and undertakes final assembly.

Specification: Boeing/Sikorsky RAH-66 Comanche (provisional)
Powerplant: two LHTEC T800-LHT-800 turboshafts each rated at 1002 kW (1,344 shp)
Dimensions: main rotor diameter 11.90 m (39 ft 0.5 in); length overall, rotor turning 14.28 m (46 ft 10.25 in) and fuselage 13.22 m (43 ft 4.5 in) excluding gun barrel; height overall 3.39 m (11 ft 1.5 in) over stabiliser; main rotor disc area 111.21 m² (1,197.14 sq ft)
Weights: empty combat equipped (including Longbow) 4167 kg (9,187 lb); normal take-off 4587 kg (10,112 lb)
Performance: maximum level speed 'clean' at optimum altitude 177 kt (328 km/h; 204 mph); maximum vertical rate of climb at sea level 360 m (1,182 ft) per minute; ferry range 1,260 nm (2335 km; 1,451 miles) with external fuel
Armament: one General Electric/GIAT two-barrelled 20-mm cannon with up to 500 rounds, plus up to six Hellfire ATGMs or Stinger AAMs in side weapons bays, and four additional Hellfires or eight Stingers, or fuel tanks on each optional detachable stub wing

Boeing Vertol CH-46 Sea Knight

US Navy UH-46Ds are employed primarily to resupply ships at sea, while HH-46Ds perform search and rescue tasks.

Still the backbone of the USMC medium assault helicopter fleet, the **CH-46 Sea Knight** is derived from the commercial Vertol Model 107 twin rotor helicopter which first flew in April 1958. The initial production variant was the **CH-46A** (160 built), which entered USMC service in June 1964. The USN also received 14 **UH-46A**s for vertical replenishment tasks. The following 266 **CH-46D**s and 10 **UH-46D**s introduced more powerful T58-GE-10 engines and cambered rotor blades. The final production variant was the **CH-46F** (174 built). The current principal version is the **CH-46E**, the result of a 'Bullfrog' update programme applied to both Ds and Fs to improve safety and crashworthiness. They introduced glass-fibre rotor blades, uprated T58-GE-16 engines and additional fuel in enlarged fuselage sponsons.

US service
The CH-46E serves with 17 USMC medium assault squadrons and a training unit. The standard load is 17 troops or 15 casualty litters. Typically 12 CH-46s will deploy aboard an amphibious assault ship. Once they have been used to establish a beach-head, they will continue a shuttle between ship and shore, bringing in extra forces and supplies. CH-46s are also used in a Special Forces support role. Additional variants are the **VH-46F** VIP transport and **HH-46D** SAR aircraft. Navy Sea Knights serve with five units, flying a variety of SAR and fleet support duties.

Licence production and exports
The CH-46 was licence-built by Kawasaki in Japan as the **KV-107**. Improved variants include the JMSDF's **KV-107/II-3** mine countermeasures version, JGSDF's **KV-107/II-4** tactical cargo/troop carrier and JASDF's **KV-107/II-5** long-range SAR version. Saudi Arabia operates several Kawasaki-built variants including a **KV-170/IIA-17** long-range passenger/cargo transport, and **KV-107/IIA** sub-designated firefighters (**SM-1**), aero-medical/rescue helicopters (**SM-2**), transports (**SM-3**), and air ambulance helicopters (**SM-4**). SAR-tasked CH-46s are operated by Sweden (**Hkp 4**) and Canada (**CH-113 Labrador**). The latter are being replaced by Bell 412HPs.

The Marine Corps is the largest CH-46 operator, the CH-46E 'Bullfrog' being the service's principal assault helicopter.

Boeing Vertol CH-46E Sea Knight

Specification: Boeing Vertol CH-46E Sea Knight
Powerplant: two General Electric T58-GE-16 turboshafts each rated at 1394 kW (1,870 shp)
Dimensions: rotor diameter, each 15.24 m (50 ft 0 in); length overall, rotors turning 25.40 m (83 ft 4 in) and fuselage 13.66 m (44 ft 10 in); height 5.09 m (16 ft 8.5 in) to top of rear rotor head; rotor disc area, total 364.82 m² (3,926.99 sq ft)
Weights: empty 5255 kg (11,585 lb); maximum take-off 11022 kg (24,300 lb); maximum payload 3175 kg (7,000 lb)
Performance: maximum speed at sea level 144 kt (267 km/h; 166 mph); maximum cruising speed at sea level 143 kt (266 km/h; 165 mph); maximum rate of climb at sea level 523 m (1,715 ft) per minute; service ceiling 2865 m (9,400 ft); hovering ceiling 2895 m (9,500 ft) in ground effect and 1753 m (5,750 ft) out of ground effect; ferry range 600 nm (1112 km; 691 miles); range with 1088-kg (2,400-lb) payload 550 nm (1019 km; 633 miles)

Boeing Vertol CH-47 Chinook

The Spanish army's 18 CH-47s are all to CH-47D standard. Nine new-build aircraft serve with converted CH-47Cs.

The **Boeing Vertol CH-47 Chinook** is the US Army's standard medium-lift helicopter and utilises Vertol's proven twin-rotor concept with externally-mounted engines. The first of 354 **CH-47A**s was first flown on 21 September 1961 and the type entered service in August 1962. The following **CH-47B** had uprated engines and increased-diameter rotor blades. The **CH-47C** introduced greater improvements, including further uprated engines and additional fuel. A total of 270 was built, of which 182 were retrofitted with composite blades and crashworthy fuel systems.

Improved D model

The current variant is the improved **CH-47D**, which first flew in February 1982. This is a mix of conversions from all three former variants and some new-build machines. The full programme covers 472 aircraft with T55-L-712 turboshafts (with a greater emergency power reserve and greater battle damage resistance), a new NVG-compatible flight deck and triple cargo hooks. It can carry up to 55 troops and a wide variety of loads up to a maximum of 10341 kg (22,798 lb) externally or 6308 kg (13,907 lb) internally. The **CH-47D International Chinook (Model 414)** is an export-optimised variant. US Army re-equipment with the CH-47D is largely complete, the variant having been supplied to 17 active-duty units, and several ARNG and Reserve organisations. Foreign operators are Egypt, Greece, Iran, Italy, Japan, Libya, Morocco, the Netherlands, South Korea, Spain, Taiwan, Thailand and the UK. The RAF's **Chinook HC.Mk 2**s are equivalent to the CH-47D.

Special forces variant

The US Army's **MH-47E SOA**s (Special Operations Aircraft) are used for covert infil/exfil work. They have a fixed IFR probe, NVG-compatible advanced cockpit displays, jam-resistant communications, a terrain-following and mapping radar and FLIR. Comprehensive defences include missile-, laser- and radar-warning receivers, jammers and chaff/flare dispensers. MH-47Es are armed with two M2 12.7-mm (0.50-in) machine-guns, and Stinger AAMs.

RAF Chinook HC.Mk 1s have been upgraded to CH-47D (HC.Mk 2) standard. They served with distinction in the Gulf.

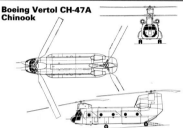

Boeing Vertol CH-47A Chinook

Specification: Boeing Vertol CH-47D Chinook
Powerplant: two Textron Lycoming T55-L-712 turboshafts each rated at 2796 kW (3,750 shp) for take-off and 2237 kW (3,000 shp) for continuous running
Dimensions: rotor diameter, each 18.29 m (60 ft 0 in); length overall, rotors turning 30.14 m (98 ft 10.75 in) and fuselage 15.54 m (51 ft 0 in); height 5.77 m (18 ft 11 in) to top of rear rotor head; rotor disc area, total 525.34 m² (5,654.86 sq ft)
Weights: empty 10151 kg (22,379 lb); maximum take-off 22679 kg (50,000 lb); maximum payload 10341 kg (22,798 lb)
Performance: maximum cruising speed at optimum altitude 256 km/h (159 mph); maximum rate of climb at sea level 669 m (2,195 ft) per minute; service ceiling 6735 m (22,100 ft); hovering ceiling 3215 m (10,550 ft) out of ground effect; operational radius between 185 and 56 km (115 and 35 miles) with maximum internal and maximum external payloads

British Aerospace Hawk

The Hawk 100 dual-role weapons trainer/ground attack aircraft has a revised 'combat wing' with wingtip missile rails, combat manoeuvre flaps, and four wing hardpoints.

The **BAe (Hawker Siddeley) Hawk T.Mk 1** trainer first flew in August 1974. It has a low-mounted wing, stepped tandem seats and is powered by a single Adour turbofan. It entered service in 1976, replacing Hunter and Gnat advanced trainers. The RAF's 175 Hawk T.Mk 1s are fitted with three weapons stations as standard for advanced tactical training. In 1983, 88 were modified to **Hawk T.Mk 1A** standard as back-up, point-defence fighters with two AIM-9L AAMs and a centreline 30-mm cannon pod.

In 1977 BAe introduced the **Series 50** upgrade which was exported to Finland, Indonesia, and Kenya. The follow-on **Series 60** introduced a 25.4-kN (5,700-lb st) Mk 861 Adour engine, and an 'advanced wing' with additional leading-edge fences and revised flaps. The Series 60 is operated by Abu Dhabi, Dubai, Kuwait, Saudi Arabia, Switzerland, South Korea (equipped with ranging radar) and Zimbabwe. The US Navy's related **T-45 Goshawk** is described separately.

BAe was able to offer the Hawk as a dedicated dual-role weapon systems trainer and ground attack aircraft. Otherwise dimensionally similar to the Hawk 60, the **BAe Hawk 100** introduced an uprated Adour Mk 871 turbofan, an increased span wing (with combat manoeuvre flaps, wingtip missile launch rails and six stores stations), a lengthened nose housing an optional FLIR and/or laser sensors, an advanced cockpit with MFDs and HOTAS, and attack-optimised avionics. A single 30-mm ADEN gun pod is an optional fitting on the fuselage centreline in place of a further stores station. First flying in October 1987, the Hawk 100 has been ordered by Abu Dhabi, Brunei, Malaysia, Oman and Saudi Arabia.

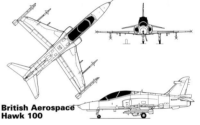

British Aerospace Hawk 100

Specification: British Aerospace Hawk Mk 100 (Enhanced Ground Attack Hawk)
Powerplant: one Rolls-Royce/Turboméca Adour Mk 871 turbofan rated at 26.00 kN (5,845 lb st)
Dimensions: wing span 9.39 m (30 ft 9.75 in) with normal tips, or 9.94 m (32 ft 7.875 in) with tip-mounted AIM-9 AAMs; length 11.68 m (38 ft 4 in) including probe; height 3.99 m (13 ft 1.25 in); wing area 16.69 m² (179.60 sq ft)
Weights: empty 4400 kg (9,700 lb); normal take-off about 5148 kg (11,350 lb); maximum take-off 9100 kg (20,061 lb)
Performance: maximum level speed 'clean' at 10975 m (36,000 ft) 560 kt (1038 km/h; 645 mph); maximum rate of climb at sea level 3597 m (11,800 ft) per minute; service ceiling 13545 m (44,500 ft); combat radius 275 nm (510 km; 317 miles) on a hi-lo-hi attack mission with seven BL755 cluster bombs
Armament: one optional centreline 30-mm ADEN cannon; maximum ordnance 3000 kg (6,614 lb) including BL 755 cluster bombs and AIM-9L AAMs

Single-seat variant

The export success of the dual-seat Hawk models led BAe to develop a single-seat **Hawk 200** variant. It retains the Hawk 100's engine, combat wing and stores-carrying capability, but has a redesigned forward fuselage for a single cockpit, an APG-66H radar in a reprofiled nose, and two 25-mm ADEN cannon. The Hawk 200 first flew in May 1986, and has been ordered by Malaysia, Oman and Saudi Arabia.

The Hawk has been a major success for the British aerospace industry. The Indonesian air force received 20 T.Mk 53s.

British Aerospace Nimrod

*The Nimrod MR.Mk 2 is the RAF's submarine hunter.
Weapons pylons (for Sidewinder self-defence AAMs) and
a refuelling probe were added for Falklands operations.*

The **BAe (HS) Nimrod** was developed from the Comet airliner as a replacement for the Avro Shackleton. Development began in 1964, when two unsold Comet 4Cs were converted to serve as prototypes with tail MAD 'stinger', nose-mounted search radar and a fin-tip ESM football. A new 14.78-m (48-ft 6-in) long ventral weapons bay was added beneath the cabin, giving a distinctive 'double-bubble' cross-section. The first Spey-powered prototype made its maiden flight on 23 May 1967. A total of 46 **Nimrod MR.Mk 1**s was ordered with the type entering service in October 1969, eventually equipping five squadrons.

Seven airframes were grossly modified to **AEW.Mk 3** standard, with Marconi radar housed in huge radomes at either end of the fuselage. The ill-fated project was cancelled in 1986 following radar development problems.

Nimrod MR.Mk 2
From 1975 the 35 remaining MR.Mk 1s were upgraded to **MR.Mk 2** configuration with a new central tactical system, (based on a new computer and three separate processors for navigation systems, radar and acoustic sensors), Searchwater radar and new communications equipment. Operation Corporate in 1982 added IFR probes and underwing weapons pylons, resulting in the modified designation **MR.Mk 2P**. All aircraft now have wingtip Loral ESM pods and can carry BOZ 100 chaff dispenser pods.

Elint R.Mk 1 platforms
Three further aircraft were ordered as **Nimrod R.Mk 1** Elint platforms to serve with No. 51 Squadron. These differ from maritime Nimrods in having no MAD tailboom and no searchlight, instead having dielectric radomes in the nose of each external wing tank and on the tail. The aircraft have been progressively modified since they were introduced, gaining additional antennas above and below the fuselage and wing tanks, as well as Loral wingtip ESM pods. With inflight-refuelling probes the designation changes to **R.Mk 1P**. For improved navigation capability, an inertial navigation system and nose-mounted weather radar were fitted.

*Nimrod MR.Mk 2s can also undertake general maritime
work such as SAR patrols and oilfield surveillance.*

**British Aerospace
Nimrod MR.Mk 2P**

Specification: British Aerospace (Hawker Siddeley) Nimrod MR.Mk 2
Powerplant: four Rolls-Royce RB.168-20 Spey Mk 250 turbofans each rated at 54.00 kN (12,140 lb st)
Dimensions: wing span 35.00 m (114 ft 10 in); length 38.63 m (126 ft 9 in); height 9.08 m (29 ft 8.5 in); wing area 197.04 m² (2,121.00 sq ft)
Weights: typical empty 39010 kg (86,000 lb); maximum normal take-off 80514 kg (177,500 lb); maximum overload take-off 87091 kg (192,000 lb)
Performance: maximum cruising speed at optimum altitude 475 kt (880 km/h; 547 mph); economical cruising speed at optimum altitude 425 kt (787 km/h; 490 mph); typical patrol speed at low level 200 kt (370 km/h; 230 mph) on two engines; service ceiling 12800 m (42,000 ft); typical endurance 12 hours, 15 hours maximum
Armament: maximum ordnance 6124 kg (13,500 lb) including AIM-9 Sidewinder AAMs for self-defence, AGM-84 Harpoon ASMs, Stingray torpedoes, bombs or depth charges

British Aerospace Sea Harrier

Identified by its bulbous radome housing Blue Vixen radar, the FRS.Mk 2 upgrade enables the Sea Harrier to remain an effective interceptor. It can carry up to four AMRAAMs.

Developed from the RAF's Harrier GR.Mk 3 STOVL strike aircraft, the **Sea Harrier FRS.Mk 1** introduced revised forward fuselage contours with a Ferranti Blue Fox radar, a revised canopy and raised cockpit for improved view, and a 96.3-kN (21,492-lb st) Pegasus Mk 104 engine. Avionics changes included addition of an auto-pilot, a revised nav/attack system and a new HUD. An initial order was placed in 1975 for 24 FRS.Mk 1s and a single **T.Mk 4A** trainer. The first operational squadron (No. 899) was commissioned in April 1980 and two squadrons (Nos 800 and 801) were subsequently deployed during the Falklands War where they served with distinction, scoring 23 confirmed victories.

Post-Falkland attrition replacements and further orders subsequently took total RN procurement up to 57 FRS.Mk 1s and four trainers (including three **T.Mk 4N**s). Improvements included revised wing pylons for carriage of four AIM-9Ls (on twin launch rails), larger-capacity drop tanks and installation of an improved Blue Fox radar and RWR. In 1978, the Indian Navy became the second Sea Harrier operator, ordering a total of 24 **FRS.Mk 51**s and four **T.Mk 60** trainers.

Improved FRS.Mk 2 variant

A mid-life update was initiated in 1985 to refine the Sea Harrier as a more capable interceptor. BAe converted two FRS.Mk 1s to serve as **FRS.Mk 2** prototypes, with the first flying in September 1988. Despite the addition of an extra equipment bay and a recontoured nose to house the Blue Vixen radar (giving compatibility with BVR AMRAAMs), the FRS.Mk 2 is actually nearly 0.61 m (2 ft) shorter overall due to the elimination of the FRS.Mk 1's pitot probe. The cockpit introduces new multi-function CRT displays and HOTAS controls. On 7 December 1988 a contract was awarded for the conversion of 31 FRS.Mk 1s to Mk 2 standard. Carrier qualification trials were conducted during November 1990. In order to enhance pilot conversion training, a new two-seat **T.Mk 8N** trainer has been mooted. Essentially a reconfigured T.Mk 4N, it would duplicate FRS.Mk 2 systems, apart from radar. An additional order for 18 further conversions was announced in January 1994.

The Fleet Air Arm's surviving Sea Harrier FRS.Mk 1s of Falklands fame are being upgraded to FRS.Mk 2 standard.

British Aerospace Sea Harrier FRS.Mk 2

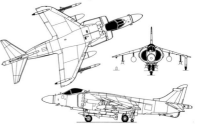

Specification: British Aerospace Sea Harrier FRS.Mk 2
Powerplant: one Rolls-Royce Pegasus Mk 106 turbofan rated at 95.64 kN (21,500 lb st)
Dimensions: wing span 7.70 m (25 ft 3 in); length 14.50 m (47 ft 4 in); height 3.71 m (12 ft 2 in); wing area 18.68 m² (201.10 sq ft)
Weights: operating empty 6374 kg (14,052 lb); maximum take-off 11884 kg (26,200 lb)
Performance: maximum level speed 'clean' at sea level more than 639 kt (1185 km/h; 736 mph); service ceiling 15545 m (51,000 ft); combat radius 400 nm (750 km; 460 miles) on a hi-hi-hi interception mission with four AAMs, or 250 nm (463 km; 288 miles) on a hi-lo-hi attack mission
Armament: two 30-mm ADEN cannon in underfuselage pods, maximum ordnance 2270 kg (5,000 lb) VTO, or 3630 kg (8,000 lb) STO, primary armament of four AIM-120 AMRAAM missiles; other stores include BL 755 and 1,000-lb (454-kg) bombs, rockets, ALARM missiles, Sea Eagle AShMs and 864-litre (190-Imp gal) drop tanks

British Aerospace VC10

During a Gulf War mission, an RAF VC10 K.Mk 2 refuels two Jaguars. The VC10's former airliner cabin is replaced by five fuel tanks, each housing 3182 litres (700 Imp gal).

Modification of the civil **Vickers/BAC VC10** airliner into a transport gave the RAF useful passenger and cargo-carrying capacity. Meeting a 1960 specification for a strategic long-range transport for the RAF's Transport Command, the first military VC10s were similar to the civil Standard VC10 but were featured uprated Conway engines, the Super VC10's additional fin fuel cell, rearward-facing seats, a side-loading freight door, an IFR probe and an APU in the tail cone. As the **VC10 C.Mk 1**, the aircraft incorporated seating capacity for up to 150 passengers or 76 stretcher cases and six medical attendents. The first of 14 RAF VC10s made its maiden flight in November 1965 and initial deliveries began in July 1966. No. 10 Squadron was the sole C.Mk 1 operator and undertook regular route flights from April 1967. Carrying less than half its full payload, the VC10 had a range exceeding 8047 km (5,000 miles).

Tanker modifications

In 1978 a programme was initiated to convert surplus VC10 airliners to tankers to augment the Victor K.Mk 2 fleet. Five Standard VC10 Series 101s and four Super VC10 Series 1154s from were converted to **VC10 K.Mk 2/3** standards respectively with extra fuel tanks in the cabin, three HDUs (two underwing and one in the rear fuselage) and a closed-circuit television system for monitoring of refuelling operations. The first VC10 K.Mk 2s joined No. 101 Squadron at Brize Norton in May 1984, with the first K.Mk 3s following in 1985. Four years later five Super VC10s were converted to short-range **VC10 K.Mk 4** standard, and eight C.Mk 1s were upgraded to **VC10 C.Mk 1(K)** standard. This retains full passenger and freight capability but introduces two underwing fuel pods. The K.Mk 4s have Mk 17 and Mk 32 IFR pods, closed-circuit TV, air-to-air TACAN, avionics systems and engines of the K.Mk 3, but no cabin fuel tanks. In 1992 it was decided to convert 13 C.Mk 1s to C.Mk 1(K) configuration. All VC10 tankers are also to be fitted with JTIDS terminals. During hostilities with Iraq, VC10s tankers were tasked to support primarily RAF strike missions; each VC10 was able to refuel a flight of four Tornados or Jaguars.

The VC10 K.Mks 2 and 3 are equipped with three refuelling hoses, all of which are trailed here.

British
Aerospace
VC10 K.Mk 3
(K.Mk 2 side
view)

Specification: British Aerospace (Vickers/BAC) VC10 C.Mk 1
Powerplant: four Rolls-Royce Conway RCo.43 Mk 301 turbofans each rated at 96.97 kN (21,800 lb st)
Dimensions: wing span 44.55 m (146 ft 2 in); length 48.38 m (158 ft 8 in) excluding probe; height 12.04 m (39 ft 6 in); wing area 272.38 m² (2,932.00 sq ft)
Weights: empty 66224 kg (146,000 lb); maximum take-off 146510 kg (323,000 lb); maximum payload 26037 kg (57,400 lb)
Performance: maximum cruising speed at 9450 m (31,000 ft) 505 kt (935 km/h; 581 mph); maximum rate of climb at sea level 930 m (3,050 ft) per minute; service ceiling 12800 m (42,000 ft); take-off distance to 10.7 m (35 ft) 2530 m (8,300 ft) at maximum take-off weight; balanced landing field length 2134 m (7,000 ft) at normal landing weight; range 3,385 nm (6273 km; 3,898 miles) with maximum payload

31

British Aerospace Harrier GR.Mk 7

Virtually all the RAF's Harrier GR.Mk 5s have now been modified to GR.Mk 7 standard, serving with four squadrons (Nos 1, 3, 4 and 20) and the SAOEU.

During the late 1970s BAe initiated independent development of an advanced Harrier. This was subsequently abandoned and the **Harrier GR.Mk 5** designation was used for a licence-built version of the **McDonnell Douglas AV-8B Harrier II** (described separately), for which BAe is a subcontractor. Two pre-series and 60 production aircraft were ordered, with the first (pre-series) flying in April 1985. Numerous detail differences from the AV-8B were specified by the RAF, with indigenous equipment such as ejection seats, self-defence systems and avionics. Problems with numerous systems (including the INS) imposed a two-year delay on RAF service entry and the aircraft were accepted lacking major equipment items, including the new 25-mm ADEN cannon, Zeus ECM system and missile approach warning system. Initial RAF deliveries began in May 1987 and the first squadron was declared operational in November 1989. Nineteen GR.Mk 5s were completed to an interim **GR.Mk 5A** standard, with provision for GR.Mk 7 avionics, and were delivered straight into storage to await conversion to full night-attack standard. Surviving GR.Mk 5s are also being converted to the same standard.

Harrier GR.Mk 7 – night-attack capability

The **Harrier GR.Mk 7** is basically the RAF equivalent of the night attack AV-8B. The redundant fairing for MIRLS is replaced by the definitive forward hemisphere antennas for the Zeus ECM system. The Harrier GR.Mk 7 also has an NVG-compatible cockpit with a digital colour map. The RAF ordered 34 GR.Mk 7s in 1988 and a converted pre-series aircraft was first flown as such in 1989. The first production GR.Mk 7 was delivered in May 1990 and operational service began in late 1992. The failure of the MIRLS recce system resulted in the GR.Mk 7 totally lacking any reconnaissance capability. Rewired GR.Mk 7s operating in Turkey were given a limited-capability with the old Harrier GR.Mk 3 camera recce pod. For the demanding conversion role, the RAF has also ordered 13 **Harrier T.Mk 10** two-seat trainers, powered by the Pegasus Mk 105 engine. The T.Mk 10 will be fitted with standard avionics, giving full combat-capablity.

A small number of Harrier GR.Mk 5s remain in service. This is a former member of No. 233 OCU, now No. 20(R) Sqn.

British Aerospace Harrier GR.Mk 7/T.Mk 10 (lower side view)

Specification: British Aerospace/McDonnell Douglas Harrier GR.Mk 7
Powerplant: one Rolls-Royce Pegasus Mk 105 turbofan rated at 96.75 kN (21,750 lb st)
Dimensions: wing span 9.25 m (30 ft 4 in); length 14.36 m (47 ft 1.5 in); height 3.55 m (11 ft 7.75 in); wing area 22.18 m² (238.70 sq ft)
Weights: operating empty 7050 kg (15,542 lb); normal take-off 10410 kg (22,950 lb) for short take-off (STO); maximum take-off 14061 kg (31,000 lb) for STO or 8595 kg (18,950 lb) for VTO
Performance: maximum level speed 'clean' at 10975 m (36,000 ft) 522 kt (967 km/h; 601 mph) and at sea level 575 kt (1065 km/h; 661 mph); take-off run 405 m (1,330 ft) at MTOW; 480 nm (889 km; 553 miles) after STO on a hi-lo-hi attack mission with seven 227-kg (500-lb) bombs
Armament: two ADEN 25-mm revolver cannon in underfuselage pods with 100 rpg; maximum ordnance 9,200 lb (4173 kg), including 500-lb and 1,000-lb bombs, BL755 cluster bombs, 68-mm SNEB rocket pods, CRV-7 rockets, CBU-87 cluster bombs, Phimat chaff/flare pod and AIM-9 AAMs

CASA C.101 Aviojet

CASA C.101 Aviojet

In the Chilean air force CASA's C.101CC-02 Aviojet has become the A-36 Halcón, an attack derivative with a distinctive 'sharp' nose and an uprated Garrett engine.

Designed by CASA with assistance from MBB and Northrop, the **C.101 Aviojet** has been built as a trainer and light strike aircraft. The first prototype made its maiden flight on 27 June 1977. Design features include an unswept wing with fixed leading edge and slotted flaps, a single Garrett TFE731 turbofan (with high bypass ratio for good fuel economy) and a stepped cockpit with tandem ejection seats. The most unusual feature is a large fuselage bay beneath the rear cockpit which can house armament or other stores. The Spanish air force purchased 88 **C.101EB-01** trainers as the **E.25 Mirlo** (Blackbird). All have hardpoints, but these are not used. A nav/attack system modernisation was introduced on all C.101EBs between 1990 and 1992.

Improved attack models

The **C.101BB** attack/trainer introduced an uprated engine and was exported as the **C.101BB-02** to Chile. Twelve **T-36**s (including eight built by ENAER) were modified with radar rangefinders in a sharper, reprofiled nose and serve as tactical weapons trainers. Four similar **C.101BB-03**s were delivered to Honduras. The dedicated attack **C.101CC** first flew on 16 November 1983 and is powered by an uprated engine with military power reserve. Twenty **C.101CC-02**s were ordered by Chile as the **A-36 Halcón** (Hawk). Most were assembled locally by ENAER and serve alongside T-36s with 1 and 12 Grupo. The first A-36 served as a demonstrator for the proposed **A-36M**, with dummy Sea Eagle missiles underwing, but this project floundered. Jordan received 16 **C.101CC-04**s for advanced training.

Attack-optimised C.101DD

In May 1985 CASA flew the prototype Dash 5-1J-engined **C.101DD**. This attack-optimised model introduces a Doppler, an inertial platform and weapon-aiming computer, and a Ferranti HUD. It also has HOTAS controls, an ALR-66 RWR and a Vinten chaff/flare dispenser, and is compatible with Maverick missiles. Intended as an improved trainer and light strike aircraft, it has yet to attract any orders.

Unsurprisingly, Spain is the chief operator of the sturdy Aviojet, flying it as a primary jet and weapons trainer. It is also the mount of the national aerobatic team, 'Team Aguila'.

Specification: CASA C.101CC Aviojet
Powerplant: one Garrett TFE731-5-1J turbofan rated at 19.13 kN (4,300 lb st) normal and 20.91 kN (4,700 lb st) with military power reserve
Dimensions: wing span 10.60 m (34 ft 9.375 in); length 12.50 m (41 ft 0 in); height 4.25 m (13 ft 11.25 in); wing area 20.00 m² (215.29 sq ft)
Weights: empty equipped 3500 kg (7,716 lb); maximum take-off 6300 kg (13,889 lb)
Performance: maximum level speed 'clean' at 6095 m (20,000 ft) 435 kt (806 km/h; 501 mph); maximum rate of climb at sea level 1494 m (4,900 ft) per minute; service ceiling 12800 m (42,000 ft); combat radius 280 nm (519 km; 322 miles) on a lo-lo interdiction mission with one cannon pod and four 250-kg (551-lb) bombs
Armament: provision for a recce pack, ECM, laser designator or twin Browning M3 12.7-mm (0.50-in) machine-gun pack with 220 rpg in underfuselage bay; maximum ordnance 2250 kg (4,960 lb) typically including one centreline DEFA 553 30-mm cannon pod, BR 250 250-kg bombs, LAU-10 127-mm (5-in) rocket pods, two AGM-65 Maverick ASMs or AIM-9L or Magic AAMs

USA

Cessna T-37 Tweet/A-37 Dragonfly
Trainer/light attack

Several nations, particularly in Central and Southern America, still operate the Cessna A-37 as a front-line attack aircraft. This is a pair of Chilean A-37B Dragonflies.

The **Cessna T-37 Tweet** was developed to meet a 1952 USAF requirement for a primary jet trainer. The aircraft has a conventional overall configuration with student pilot (at left) and instructor (at right). The first of two **XT-37** prototypes was flown on 12 October 1954, powered by YJ69-T-9s (licence-built versions of the Turboméca Marboré turbojet). An initial batch of 10 **T-37A**s was followed by a further 524 A models. During 1959 production switched to the 4.56-kN (1,025-lb st) J69-T-25-engined **T-37B model**, which also introduced improved equipment and provision for wingtip fuel tanks. A total of 466 was built, some being exported. All surviving T-37As were also brought up to T-37B standard. The T-37 was to have been replaced by the Fairchild T-46A, but this was cancelled in 1986. From 1989, the Sabreliner Corp. began supplying modification kits to the USAF to allow the T-37 to be structurally rebuilt for extended service. The ultimate **T-37C** model introduced a light attack capability; 269 were built solely for export.

Attack Tweet – the A-37 Dragonfly
A light attack derivative of the T-37 was flown in prototype form on 22 October 1963. Thirty-nine T-37Bs were similarly converted to become **A-37A**s. All aircraft featured strengthened airframe, J85 turbojets, armour protection, a 7.62-mm Minigun, eight wing hardpoints, wingtip tanks and ground-attack avionics. A full-production version was ordered by the USAF as the **A-37B Dragonfly,** introducing IFR capability and uprated engines. Deliveries of the Dragonfly began in May 1968, with a total of 577 examples delivered until 1975. The A-37B served in the Vietnam War, primarily in the close support role. At least 130 were retrospectively fitted with avionics optimised for the forward air control mission as **OA-37B**s. These were retired from US service in 1992. Both A-37B and OA-37B serve extensively with Latin American air arms in operational training and light attack roles. Operators comprise Chile, Colombia, Ecuador, Guatemala, Honduras, Peru, Salvador and Uruguay. Other current A-37B operators are South Korea and Thailand.

Until the introduction of the final JPATS winner, the T-37 'Tweet' (progressively SLEPed by the Sabreliner Corp.) will be the USAF's primary jet trainer.

Cessna A-37 Dragonfly

Specification: Cessna Model 318E (OA-37B Dragonfly)
Powerplant: two General Electric J85-GE-17A turbojets each rated at 12.68 kN (2,850 lb st)
Dimensions: wing span 10.93 m (35 ft 10.5 in) with tip tanks; length excluding probe 8.93 m (29 ft 3.5 in); height 2.70 m (8 ft 10.5 in); wing area 17.09 m² (183.9 sq ft)
Weights: basic empty 2817 kg (6,211lb); empty equipped 2650 kg (5,843 lb); maximum take-off 6350 kg (14,000 lb)
Performance: maximum level speed at 4875 m (16,000 ft) 440 kt (816 km/h; 507 mph); maximum cruising speed at 7620 m (25,000 ft) 787 km/h (489 mph); maximum rate of climb at sea level 2130 m (6,990 ft) per minute; service ceiling 12730 m (41,765 ft); range with maximum warload 399 nm (740 km; 460 miles)
Armament: one 7.62-mm (0.3-in) GAU-2B/A Minigun in forward fuselage; eight underwing pylons rated up to total of 1860 kg (4,100 lb); ordnance includes 227-kg (500-lb) Mk 82 bombs, fuel tanks, rocket launchers and gun pods

Chengdu J-7/F-7

China has worked hard to sell the F-7M Airguard overseas and has met with some success. Pakistan is a major operator, receiving over 80 similar F-7P Skybolts.

China was granted a licence to manufacture the MiG-21F-13 and its Tumanskii R-11F-300 engine in 1961. A prototype was constructed at Shenyang and first flew on 17 January 1966, powered by the Wopen WP-7 engine. Certified for production in June 1967, despite the enormous upheavals of the Cultural Revolution, the initial batch of aircraft was manufactured at Shenyang. Some were later delivered to Albania and Tanzania as **F-7A**s. Production transferred to Chengdu, receiving the new designation **J-7I**. In 1975 development began of the improved **J-7II** with more conventional canopy and Wopen WP-7B. Reports exist of an improved J-7II which first flew in April 1990 as the **J-7E** with a cranked arrow wing, uprated WP-7F and reported compatibility with the PL-8 AAM. The **F-7B** is a minimum-change variant, apparently with R550 Magic compatability, and has been exported to Egypt, Iraq, Sri Lanka (as the **F-7BS**), and reportedly Jordan.

Upgraded F-7M Airguard and F-7P Skybolt

The upgraded **F-7M Airguard** introduced two additional pylons and features much improved avionics systems (including a new longer-range ranging radar) and an improved Wopen WP-7B(BM). The F-7M has been exported to Bangladesh, Iran, Jordan (reportedly) and Zimbabwe (these may also be F-7Bs). A similar version, the **F-7P Skybolt**, introduced Western equipment such as Martin-Baker Mk 10L ejection seats and provision for AIM-9 and MATRA Magic AAMs. Pakistan received 20 F-7Ps and 60 improved **F-7MP**s with fintip-mounted RWR.

All-weather variant and further improved models

The longer-range all-weather **J-7III** is externally similar to the Soviet MiG-21M. Powered by a more reliable Wopen WP-13 engine, it has an all-weather radar in an enlarged radome. The unbuilt **Chengdu/Grumman Sabre** or **Super 7** was to have featured an APG-66 radar in a solid nosecone, improved avionics and lateral intakes. Its development was suspended after the Tienanmen Square massacre.

Guizhou has also developed a two-seat version of the F-7, designated FT-7. It can be flown armed with a ventral 23-mm cannon and underwing stores (including AAMs).

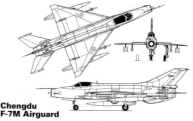

Chengdu F-7M Airguard

Specification: Chengdu F-7M Airguard
Powerplant: one Liyang (LMC) Wopen-7B(BM) turbojet rated at 43.15 kN (9,700 lb st) dry and 59.82 kN (13,448 lb st) with afterburning
Dimensions: wing span 7.15 m (23 ft 5.625 in); length 13.95 m (45 ft 9 in) excluding probe and 14.89 m (48 ft 10 in) including probe; height 4.10 m (13 ft 5.5 in); wing area 23.00 m² (247.58 sq ft)
Weights: empty 5275 kg (11,629 lb); normal take-off 7531 kg (16,603 lb)
Performance: maximum level speed 'clean' at high altitude 1,175 kt (2175 km/h; 1,350 mph); maximum rate of climb at sea level 10800 m (35,433 ft) per minute; service ceiling 18200 m (59,700 ft); combat radius 324 nm (600 km; 373 miles) on a hi-lo-hi interdiction mission with two 150-kg (331-lb) bombs and three 500-litre (132-US gal) drop tanks
Armament: two internal 30-mm cannon with 60 rpg; plus up to 1000 kg (2,205 lb) of ordnance including two PL-2, -2A, -5B or -7 AAMs (or MATRA Magic AAMs); 18-round 57-mm or seven-round 90-mm rocket pods; 50-, 150-, 250- or 500-kg bombs; or 500-litre drop tanks

Dassault Mirage III/5/50

Several air forces have upgraded their Mirage IIIs, 5s and 50s. Brazil's F-103E/Ds (10 IIIEBRs and two DBRs) were upgraded from 1989 and serve alongside five further ex-French Mirage IIIEs and two IIIBEs, which were updated by Dassault.

The classic **Dassault Mirage III** Mach 2 delta was flown in prototype form on 17 December 1956. Production of 1,422 Mirage IIIs, 5s and 50s lasted until 1992. Most early **IIIC** interceptors have been withdrawn, as have most **IIIB** trainers, although a few **IIIB-1** testbeds and **IIB-RV** refuelling trainers remain. Argentina retains 19 ex-Israeli **IIICJ** fighters. The multi-role **Mirage IIIE** fighter (and equivalent **IIID** trainer) was flown in 1961 and introduced provision for an AN52 nuclear bomb. France received 183 (plus 20 equivalent **IIIBE** trainers) which remained in defence-suppression and conventional attack roles until late 1993. The AA also received 70 **Mirage IIIR**s with camera nose (including 20 **IIRD**s with Doppler). All have been supplanted by Mirage F1CRs. The IIID/E achieved substantial export success with deliveries to Argentina, Australia, Lebanon, Pakistan and Spain. Brazil, Switzerland and Venezuela have undertaken extensive upgrade programmes, involving structural rework, addition of canards and improved avionics.

Mirage 5

A simplified non-radar ground-attack IIIE version was developed in 1966 as the **Mirage 5**. Israel's 50 embargoed **Mirage 5J**s were redelivered to the AA as **Mirage 5F**s. They were retired in late 1993. Related variants are the two-seat **Mirage 5D** trainer and recce **Mirage 5R**. The original simplified aircraft subsequently became available with an ever-greater range of avionics options, including the reintroduction of lightweight radars. The **Mirage 50** first flew in April 1979 and introduced the F1's Atar 9K-50 engine endowing better field performance, faster acceleration, larger weapon load and improved manoeuvrability.

Exports and upgrades

Mirage 5s were exported to Abu Dhabi, Gabon and Zaïre. Updated Mirage 5/50s are operated by Argentina, Belgium, Chile, Colombia, Egypt, Pakistan, Peru and Venezuela. Modifications include structural rework and addition of IFR probes, canards and improved radar and nav/attack avionics.

Venezuela's two surviving Mirage 5DV trainers have been upgraded to 50DV standard with canards and an IFR probe.

Dassault Mirage 5F/50C (side view)

Specification: Dassault Aviation Mirage 50M
Powerplant: one SNECMA Atar 9K-50 turbojet rated at 49.03 kN (11,023 lb st) dry and 70.82 kN (15,873 lb st) with afterburning
Dimensions: wing span 8.22 m (26 ft 11.6 in); length 15.56 m (51 ft 0.6 in); height 4.50 m (14 ft 9 in); wing area 35.00 m² (376.75 sq ft); optional canard foreplane area 1.00 m² (10.76 sq ft)
Weights: empty equipped 7150 kg (15,763 lb); normal take-off 10000 kg (22,046 lb); maximum take-off 14700 kg (32,407 lb)
Performance: maximum level speed 'clean' at 12000 m (39,370 ft) 1,262 kt (2338 km/h; 1,453 mph); maximum rate of climb at sea level 11160 m (36,614 ft) per minute; climb to 13715 m (45,000 ft) in 4 minutes 42 seconds; service ceiling 18000 m (59,055 ft); combat radius 710 nm (1315 km; 817 miles) of a hi-hi-hi interception mission with two AAMs and three drop tanks
Armament: two internal DEFA 552A 30-mm cannon with 125 rpg, plus up to 4000 kg (8,818 lb) of ordnance, including R 530 AAMs, 454-kg bombs, rockets, combined tanks/rocket launchers

Dassault Mirage IV

The Mirage IV was designed with a rough field capability for use from dispersed operating locations. Under such conditions, rocket-assisted take-offs are most useful.

In 1954 France created a national Force de Frappe (nuclear deterrent), one element of which would be a manned bomber. Originally planned as a bigger aircraft, the **Dassault Mirage IV** was finally scaled down as the **Mirage IVA** with two Atar turbojets, which meant that it could not fly two-way missions to Soviet targets. The prototype flew on 17 June 1959 and, following considerable further development, a total of 62 Mirage IVs was manufactured, completing the force in March 1968.

Configuration

The Mirage IVA is broadly similar to a scaled-up Mirage III, with side-by-side engines, tandem cockpits, four-wheel bogie main undercarriage and nose-mounted inflight-refuelling probe. Navigation is by surveillance radar under the belly, Doppler, central computer and autopilot, more recently upgraded by adding dual inertial systems. Strike cameras are fitted, and the original weapon load comprised a 60-kT nuclear bomb recessed into the rear fuselage, though by removing the large fuel tanks it is possible to carry six conventional bombs or four AS37 Martel ARMs. For many years the force at readiness comprised 36 aircraft (of 51 bomber versions available), dispersed in small groups. In emergency, further, small field, dispersal is possible. Some were configured for the reconnaissance mission with a semi-recessed camera/SEAR package.

Upgraded Mirage IVP

In the late 1980s 19 Mirage IVAs were converted to **Mirage IVP** standard. Together with updated avionics including a twin inertial platform and Serval RWRs, the IVP introduced the ability to carry the ASMP stand-off nuclear missile on a pylon under the belly. A reduction in the Mirage IV force means that only 14 are operational, serving with two squadrons of Escadre de Bombardement Stratégique 91. They comprise EB1/91 and EB2/91, headquartered at Mont-de-Marsan Cazaux, although Mirage IVs are deployed in four detachments for security reasons.

The ASMP (Air-Sol Moyenne Portée) missile offers a 300-kT TN81 warhead, and variable delivery profiles up to 250 km (155 miles) at high level, increasing survivability.

Dassault Mirage IVP

Specification: Dassault Aviation (Dassault-Breguet) Mirage IVP
Powerplant: two SNECMA Atar 9K-50 turbojets each rated at 49.03 kN (11,023 lb st) dry and 70.61 kN (15,873 lb st) with afterburning
Dimensions: wing span 11.85 m (38 ft 10.5 in); length 23.50 m (77 ft 1.2 in); height 5.65 m (18 ft 6.4 in); wing area 78.00 m² (839.61 sq ft)
Weights: empty equipped 14500 kg (31,966 lb); maximum take-off 31600 kg (69,666 lb)
Performance: maximum level speed 'clean' at 11000 m (36,090 ft) 1,262 kt (2338 km/h; 1,453 mph) or at low level about 728 kt (1349 km/h; 838 mph); normal penetration speed at 11000 m (36,090 ft) 1,172 kt (1913 km/h; 1,189 mph); climb to 11000 m (36,090 ft) in 4 minutes 15 seconds; service ceiling 20000 m (65,615 ft); ferry range 2,158 nm (4000 km; 2,486 miles) with drop tanks; typical combat radius 668 nm (1,240 km; 771 miles)
Armament: one 900 kg (1,984 lb) ASMP stand-off nuclear missile or maximum ordnance of 7200 kg (15,873 lb) in conventional bomber role

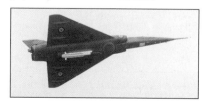

Dassault Mirage F1

France
Multi-role fighter-bomber

During Operation Daguet in 1991, the Armée de l'Air's newly modified F1CR reconnaissance aircraft displayed their secondary ground attack role by bombing Iraqi positions.

Despite its suffix, the **Mirage F1C** was the initial production version of Dassault's successor to its highly successful Mirage III/5 delta. It was developed to meet an Armée de l'Air (AA) requirement for an all-weather interceptor. Forsaking the Mirage III's delta configuration for a high-mounted wing and conventional tail surfaces, the prototype first flew on 23 December 1966. F1C production deliveries to the AA began in May 1973. The initial 83 aircraft were followed by 79 **F1C-200**s with fixed refuelling probes (necessitating a small fuselage plug). The AA also recieved 20 **F1B** tandem-seat trainers which retain full combat capability. Dassault has delivered 64 **F1CR-200** dedicated tactical reconnaissance platforms with infra-red linescan unit, undernose camera and centreline pods for SLAR, LOROP or Elint equipment. The **Mirage F1CT** (T-Tactique) was a logical product of the shortfall in French ground attack capability and a surplus of air defence fighters following Mirage 2000C deliveries. From 1991, 57 F1C interceptors are being given expanded tactical capability with a laser rangefinder, improved RWR and chaff/flare dispensers.

Export variants

Emulating the successor of its predecessor, the Mirage III/5/50, the F1 has been widely exported. F1Cs were sold to South Africa (**F1CZ**), Morocco (**F1CH**), Jordan (**F1CJ**), Kuwait (**F1CK, F1CK2**), Greece (**F1CG**) and Spain (**F1CE**). The **F1A** is a simplified version for day visual attack missions with Aïda II ranging radar. It was sold to Libya and South Africa (**F1AZ** fitted with a laser-ranger). The **F1E** (and corresponding **F1D** trainer) designation applies to an upgraded multi-role fighter/attack version for export customers with an INS, central nav/attack computer and CRT head-up display. The F1E was exported to Ecuador (**F1JA/E**), Iraq (**F1EQ**), Jordan (**F1EJ**), Libya (**F1ED**), Morocco (**F1EH and F1EH-200**), Qatar (**F1EDA and F1DDA**) and Spain (**F1EE-200**). The Iraqi **F1EQ-5** and **F1EQ-6**s were fitted with Agave radar to give an anti-shipping capability with Exocet ASMs. They were used extensively during the Persian Gulf 'tanker war'.

Iraq's F1EQ-5s and EQ-6s were equipped with Agave radar, giving compatibilty with Exocet anti-ship missiles. The slate grey camouflage was adopted for overwater operations.

Dassault Mirage F1C-200

Specification: Dassault Aviation Mirage F1C
Powerplant: one SNECMA Atar 9K-50 turbojet rated at 49.03 kN (11,023 lb st) dry and 70.21 kN (15,785 lb st) with afterburning
Dimensions: wing span 9.32 m (30 ft 6.75 in) with tip-mounted Magic AAMs; length 15.30 m (50 ft 2.5 in); height 4.50 m (14 ft 9 in); wing area 25.00 m² (269.11 sq ft)
Weights: empty 7400 kg (16,314 lb); normal take-off 10900 kg (24,030 lb); maximum take-off 16200 kg (35,715 lb)
Performance: maximum level speed 'clean' at 11000 m (36,090 ft) 1,262 kt (2338 km/h; 1,453 mph); maximum rate of climb at sea level 12780 m (41,930 ft) per minute with afterburning; service ceiling 20000 m (65,615 ft); combat radius 229 nm (425 km; 264 miles) on a hi-lo-hi attack mission with 14 250-kg (551-lb) bombs
Armament: two internal DEFA 553 30-mm cannon with 135 rpg; (intercept) Super 530 and R550 Magic AAMs; maximum ordnance of 6300 kg (13,889 lb) including LGBs, ASMs and bombs

Dassault Mirage 2000

For export, Dassault has developed the Mirage 2000-5 in both single- and two-seat versions. Taiwan was the launch customer, ordering 60 aircraft, along with 1,000 Magic 2 and MICA AAMs.

For the third Mirage generation, Dassault returned to the delta configuration, using negative longitudinal stability and a fly-by-wire flight control system to eliminate many of the shortcomings of a conventional delta. As such, the **Mirage 2000** has its predecessor's large internal volume and low wave drag, but has improved agility, slow-speed handling and lower landing speed. The first of five prototypes was initially flown on 10 March 1978. The first of 37 production **Mirage 2000C**s made its maiden flight on 20 November 1982 and 2000C deliveries began in April 1983, with IOC in July 1984. All early production aircraft have M53-5 engines and the multi-role RDM radar. Later 2000Cs introduced M53-P2 powerplants and RDI radar optimised for look-down/shoot-down intercepts with two MATRA Super 530D missiles. With RDM radar, Mirage 2000Cs carried Super 530F and Magic 1 missiles.

Variants and exports

The **Mirage 2000B** tandem two-seat trainer first flew in August 1983. It loses some internal fuel and both cannon in order to accommodate the second cockpit. The recce **Mirage 2000R** has a radar nose and carries podded sensors: multi-camera, side-looking airborne radar, and long-range optical. For export, the RDM-equipped, M53-P2-powered variant is designated **Mirage 2000E/ED/ER**. These have been delivered to Abu Dhabi, Egypt, Greece, India and Peru.

Mirage 2000-5 upgrade

The upgraded **Mirage 2000-5** introduces the Rafale's five-screen cockpit display, MICA AAMs, RDY multi-mode radar and additional avionics. A trainer prototype flew on 24 October 1990. Options available include Super 530 or Sky Flash AAMs in place of MICA and (from 1995) a 98.06-kN (22,046-lb st) M53-P20. The 2000-5 is also able to launch AM39 Exocet ASMs as well as ARMAT, AS30L and LGBs. The APACHE stand-off weapons dispenser is to be integrated when it enters service. Taiwan placed an order for 60 aircraft in late 1992 for initial delivery in 1995. Some French 2000Cs are to be upgraded to this standard from 1994-97.

Greece is the only European Mirage 2000 export customer, operating Mirage 2000EGs and 2000BGs alongside the F-16.

Dassault Mirage 2000B

Specification: Dassault Aviation Mirage 2000C
Powerplant: one SNECMA M53-P2 turbofan rated at 64.33 kN (14,462 lb st) dry and 95.12 kN (21,384 lb st) with afterburning
Dimensions: wing span 9.13 m (29 ft 11.5 in); length 14.36 m (47 ft 1.25 in); height 5.20 m (17 ft 0.75 in); wing area 41.00 m^2 (441.33 sq ft)
Weights: empty 7500 kg (16,534 lb); normal take-off 10680 kg (23,545 lb); maximum take-off 17000 kg (37,478 lb)
Performance: maximum level speed 'clean' at 11000 m (36,090 ft) more than 1,262 kt (2338 km/h; 1,453 mph); maximum rate of climb at sea level 17060 m (55,971 ft) per minute; service ceiling 18000 m (59,055 ft); combat range over 800 nm (1480 km; 920 miles) with four 250-kg bombs
Armament: two internal 30-mm DEFA 554 cannon with 125 rpg; (intercept) two Super 530D or 530F and two Magic 2 AAMs; (attack) up to 6300 kg (13,889 lb) of ordnance, including 250-kg retarded, BAP 100 anti-runway, cluster bombs and 1000-kg LGBs, AS30L and ARMAT anti-radar missiles, rockets and fuel tanks

Dassault Mirage 2000D/N/S

The Mirage 2000N will replace the Mirage IVP as the French air force's primary strike aircraft. The specially-developed 2000-litre (440-Imp gal) tanks have unusual bulbous front ends.

French requirements for a Mirage IVP replacement to carry the ASMP stand-off nuclear bomb resulted in Dassault receiving a contract in 1979 for two **Mirage 2000P (Pénétration)** prototypes (later designated **2000N (Nucléaire)**). Based on the 2000B two-seat trainer, it has a strengthened airframe for low-level flight and considerable differences in avionics, including twin INSs, and Antilope 5 radar optimised for terrain following, ground mapping and navigation. It provides automatic terrain following down to 91 m (300 ft) at speeds up to 600 kt (1112 km/h; 691 mph). Both pilot and WSO have moving map displays. ASMP delivers a 150- or 300-kT warhead up to 80 km (50 miles) from a low-altitude launch point. Outboard, the Mirage 2000N carries a pair of large, 2000-litre (440-Imp gal) drop tanks and two self-defence MATRA Magic AAMs. Further protection is provided by the Serval RWR, Sabre electronic jammers and a Spirale chaff/flare system. From the 32nd French 2000N onwards, Mirage 2000Ns are equipped to carry alternative loads of conventional ordnance up to a maximum of 6300 kg (13,890 lb).

Deployment

Initially flown on 3 February 1983, the 2000N achieved IOC in July 1988 when the first of an eventual three squadrons was reformed. Orders were decreased, but delays with the Rafale generated a requirement for more aircraft with conventional weapons capability. The latter became the **Mirage 2000N' (N Prime)**, a confusing designation amended to **2000D**. When further orders were curtailed in 1991, France had ordered 75 Mirage 2000Ns (31 ASMP, 44 ASMP/ conventional) and 75 Mirage 2000Ds. The first 2000D flew on 19 February 1991, its differences from earlier standard comprising deletion of the interface between ASMP and the aircraft's navigation equipment, addition of a GPS and a redesign of cockpit instrumentation. All versions have provision for an IFR probe immediately ahead of the windscreen. An export **Mirage 2000S (Strike)** version is to be available from 1994 retaining Antilope 5 radar and terrain-following capability. No orders have been received.

The Mirage 2000S is a non-nuclear export derivative of the 2000D. This example carries an ATLIS laser designator pod, an AS30L laser-guided missile and a MATRA Magic 2 AAM.

Dassault Mirage 2000N

Specification: Dassault Aviation Mirage 2000C
Powerplant: one SNECMA M53-P2 turbofan rated at 64.33 kN (14,462 lb st) dry and 95.12 kN (21,384 lb st) with afterburning
Dimensions: wing span 9.26 m (30 ft 4.5 in); length 14.55 m (47 ft 9 in); height 5.15 m (16 ft 10.75 in); wing area 41.00 m² (441.33 sq ft)
Weights: empty 7500 kg (16,534 lb); maximum take-off 17000 kg (37,478 lb)
Performance: maximum level speed 'clean' at 11000 m (36,090 ft) more than 1,262 kt (2338 km/h; 1,453 mph); penetration speed at 60 m (197 ft) 600 kt (1112 km/h; 691 mph); maximum rate of climb at sea level 17060 m (55,971 ft) per minute; service ceiling 18000 m (59,055 ft); combat range more than 800 nm (1480 km; 920 miles)
Armament: (alternative to ASMP) up to 6300 kg (13,890 lb) of ordnance including AS30L missiles and BGL bombs, APACHE munitions dispensers, AM39 Exocet ASMs, ARMAT ARMs, Durandal anti-runway bombs and other rockets, cluster bombs and area-denial weapons

Dassault Rafale

The Rafale embodies France's determination to develop its own advanced combat aircraft. This view illustrates the three Rafale variants (black-painted C, two-seat B and marine M).

Dassault Rafale C

Dassault's **Avion de Combat Experimentale**, or **ACX**, evolved as an early 1980s technology demonstrator for a national combat aircraft programme even before France's withdrawal from the EFA project in August 1985. The **Rafale A** ACX demonstrator was first flown on 4 July 1986 and established and proved the basic design, configuration and performance of the planned **Rafale**, or **ACT (Avion de Combat Tactique)**, as well as its fly-by-wire control system and mainly composite structure, although using two GE F404s as interim powerplants. The port engine was replaced by an M88 powerplant in February 1990.

Rafale variants

The Armée de l'Air's generic **Rafale D** family (*discret*, or stealth) is four per cent smaller than Rafale A, with an empty weight below 9000 kg (19,840 lb), and introduces measures to reduce radar cross-section. The **Rafale C** is the AA's single-seat multi-role combat version and a prototype first flew in May 1991. SO2 production standard aircraft will have ASMP missiles, automatic terrain-following, Spectra defensive sub-systems, helmet-mounted sight and OSF IRST system. To accelerate the programme, the first 20 S01 initial standard aircraft for each service will not have these features. The **Rafale M** is the Aéronavale's single-seat carrierborne fighter with 80 per cent structural and equipment commonality with Rafale C, and 95 per cent systems commonality. Modifications for carrier operations include an arrester hook, a 'jump strut' nosewheel leg and deletion of forward centreline pylon. The **Rafale B** was originally planned as a dual-control, two-seat variant, but is now being developed into a fully operational version. A prototype first flew in April 1993 with electronic scanning RBE2 radar and Spectra defensive systems.

More two-seaters

Following Gulf War operational experience, the AA revised its Rafale procurement plans to increase the proportion of two-seaters.Total procurement currently stands at 312 Rafales, comprising 95 Cs, 135 Ds and 78 Ms.

The Rafale A demonstrator flew as early as 1986, long before comparable programmes such as the Gripen or Eurofighter.

Specification: Dassault Aviation Rafale C
Powerplant: two SNECMA M88-3 turbofans each rated at 86.98 kN (19,555 lb st) with afterburning
Dimensions: wing span 10.90 m (35 ft 9.125 in) with tip-mounted AAMs; length 15.30 m (50 ft 2.5 in); wing area 46.00 m² (495.16 sq ft)
Weights: maximum take-off 21500 kg (47,399 lb)
Performance: maximum level speed 'clean' at 11000 m (36,090 ft) 1,147 kt (2125 km/h; 1,321 mph); combat radius 590 nm (1093 km; 679 miles) on a low-level penetration mission with stores and fuel
Armament: one 30-mm GIAT DEFA 791B cannon in port engine intake trunking; 14 stores stations for maximum of 6000 kg (13,228 lb) of ordnance including one ASMP stand-off nuclear weapon (nuclear strike); up to eight MATRA MICA AAMs (with IR or active homing) and two fuel tanks (intercept); 227-kg (500-lb) bombs, APACHE stand-off munitions dispensers, LGBs, AS30L laser ASMs and an ATLIS designator pod (air-to-surface); and two AM39 Exocet AShMs (maritime strike)

Dassault/Dornier Alpha Jet

France's advanced training requirements are undertaken by the Alpha Jet. For weapons training duties, a single 30-mm DEFA cannon pod may be carried on the centreline.

In July 1969 Dassault and Dornier agreed to jointly develop and produce a new advanced trainer. The resultant **Alpha Jet** had swept shoulder-mounted wings, two Larzac turbofans and stepped tandem cockpits. French and German equipment fits vary considerably, the Luftwaffe having then decided to continue military pilot training in the US, changing its requirements to a light ground-attack replacement for its Fiat G91R/3s. This necessitated advanced nav/attack systems, including a twin-gyro INS, Doppler navigation radar, HUD, and a belly-mounted 27-mm Mauser cannon pod (30-mm DEFA pod on French Alphas). The initial order for 200 aircraft for each country was reduced to 175.

Franco-German trainers and exports

Alpha Jet development was finally approved in February 1972, and two prototypes were flown in France and Germany in 1973 and 1974 respectively. French production **Alpha Jet E**s (Ecole) began flying in November 1977 and service trials commenced in 1978. German production started with the first **Alpha Jet A** (Appui Tactique) flying in April 1978. In 1993 Germany retired all but 20 German Alphas (for lead-in training for Tornado crews) and a total of 50 surplus aircraft was given to Portugal. Initial exports were made to Belgium (33), Egypt (30 including 26 locally-assembled **Alpha Jet MS 1** trainers), Ivory Coast (12), Morocco (24), Nigeria (24), Qatar (six) and Togo (five).

Nouvelle Génération

For lead-in fighter training and light ground-attack, Dassault launched the **Alpha Jet NGEA** (Nouvelle Génération Appui/Ecole) or **MS2** programme in 1980. It featured uprated engines and new avionics including an INS, CRT HUD and laser rangefinder, plus provision for Magic AAMs. Customers were Cameroon (seven) and Egypt (15). An MS2-derived **Alpha Jet 3 Advanced Training System**, or **Lancier**, with twin multi-function cockpit displays for mission training with such sensors as AGAVE or Anemone radar, FLIR, laser, video and ECM systems, plus advanced weapons. This has yet to proceed beyond testbed stage.

The Luftwaffe's Alpha Jets were initially used for attack duties, but the survivors are now used only for training.

Dassault Alpha Jet A

Specification: Dassault Aviation (Dassault-Breguet)/Dornier Alpha Jet E (Alpha Jet Advanced Trainer/Light Attack Version)
Powerplant: two SNECMA/Turboméca Larzac 04-C6 turbofans each rated at 13.24 kN (2,976 lb st)
Dimensions: wing span 9.11 m (29 ft 10.75 in); length 11.75 m (38 ft 6.5 in); height 4.19 m (13 ft 9 in); wing area 17.50 m² (188.37 sq ft)
Weights: empty equipped 3345 kg (7,374 lb); normal take-off 5000 kg (11.023 lb); maximum take-off 8000 kg (17,637 lb)
Performance: maximum level speed 'clean' at sea level 539 kt (1000 km/h; 621 mph); maximum rate of climb at sea level 3660 m (12,008 ft) per minute; service ceiling 14630 m (48,000 ft); operational radius 361 nm (670 km; 416 miles) on a lo-lo-lo training mission with two drop tanks; endurance more than 3 hours 30 minutes at high altitude on internal fuel
Armament: one ventral cannon pod (27-mm Mauser or 30-mm DEFA), plus four underwing stations for up to 2500 kg (5,511 lb) of stores, including bombs, rockets, missiles or drop tanks

EH Industries EH.101 Merlin

United Kingdom/Italy
Multi-role helicopter

The ASW Merlin HAS.Mk 1 is the intended replacement for the Royal Navy's Sea Kings, and will typically operate from Type 23 frigates and 'Invincible'-class aircraft-carriers.

Ranking as one of Europe's most important current helicopter programmes, the **EH.101** has its roots in the cancelled Westland WG 34 design that was adopted in late 1978 to replace the Sea King. Negotiations between Westland and Agusta in November 1979 led to the establishment of European Helicopter Industries Ltd to manage the programme. Although the Sea King replacement was the primary objective, several other potential roles were planned from the outset, including military and civil transport and utility duties. For some of these, the rear fuselage was to be modified with a loading ramp.

Advanced technology

The EH.101 is a three-engined helicopter with a single five-bladed composite main rotor with BERP-derived high-speed tips. Composite materials are also used extensively throughout the airframe. Systems and equipment vary with role and customer. IBM is the prime contractor in association with Westland for the Royal Navy's RTM 322-powered **Merlin HAS.Mk 1** and provides some equipment as well as overall management; other avionics include Blue Kestrel 360° search radar, AQS-903 processing and display system, Orange Reaper ESM and dipping sonar. The initial RN requirement for 50 Merlins has been reduced to 44 for delivery starting in 1996. The Italian navy is expected to acquire up to 24 GE T700-powered EH.101s. Earlier variants of the GE engine were used to power the nine prototypes (five built by Westland, four by Agusta). The first of these flew in England in October 1987 and all had flown by April 1991. A military utility variant first flew in December 1989, and is expected to be ordered by the RAF as its next logistic and tactical support medium-lift helicopter.

Cancelled Canadian orders

The first customer for the utility variant was Canada, which ordered 15 utility **CH-149 Chimo**s for SAR duties, along with 35 **CH-148 Petrel** naval versions to meet its New Shipborne Aircraft requirement. However, this order was subsequently cancelled in late 1993.

The military utility EH.101 can lift 6 tonnes or 30 troops. This is PP9, the final, Agusta-built (civil) utility version.

EH Industries
EH.101 Merlin

Specification: European Helicopter Industries EH.101 Merlin (naval model)
Powerplant: three Rolls-Royce/Turboméca RTM322-01 turboshafts each rated at 1724 kW (2,312 shp) and 1566 kW (2,100 shp) for maximum and intermediate contingencies respectively
Dimensions: main rotor diameter 18.59 m (61 ft 0 in); tail rotor diameter 4.01 m (13 ft 2 in); length overall, rotors turning 22.81 m (74 ft 10 in) and fuselage 22.80 m (74 ft 9.6 in); height overall 6.65 m (21 ft 10 in) with rotors turning; main rotor disc area 271.51 m² (2,922.60 sq ft)
Weights: (estimated) basic empty 7121 kg (15,700 lb); operating empty 9298 kg (20,500 lb); maximum take-off 13530 kg (29,830 lb)
Performance: average cruising speed 160 kt (296 km/h; 184 mph); ferry range 1,000 nm (1853 km; 1,152 miles) with auxiliary fuel; endurance 5 hours on station with maximum weapon load
Armament: maximum ordnance 960 kg (2,116 lb) comprising four Sting Ray homing torpedoes plus two sonobuoy dispensers; optional Exocet, Harpoon, Sea Eagle and Marte Mk 2 AShMs

EMBRAER/Shorts Tucano

EMBRAER has teamed with Northrop to offer the improved EMB-312H Tucano for the US JPATS trainer requirement. This features a stretched fuselage and an uprated PT6 engine.

Development of the **EMBRAER EMB-312 Tucano** (Toucan) high-performance turboprop trainer started in 1978 in response to a Brazilian air force specification for a Cessna T-37 replacement. First flown on 16 August 1980, the first of 133 **T-27**s ordered (plus 40 options) entered service in September 1983. Designed from the outset to provide a 'jet-like' flying experience, the Tucano has a single control lever governing both propeller pitch and engine throttling, ejection seats, and a staggered tandem-place cockpit. Four underwing hardpoints can carry ordnance for weapons training. Tucanos have been exported to Argentina (30), Colombia (14), Egypt (54), France (80), Honduras (12), Iran (25), Iraq (84), Paraguay (six), Peru (30) and Venezuela (31). In June 1991, EMBRAER announced development of the **EMB-312H**, featuring an uprated 1193-kW (1,600-shp) P&WC PT6A-68/1 engine and a stretched fuselage. This version is being bid in the USAF/USN JPATS competition.

Shorts Tucano

The Tucano's most notable export success came in March 1985, when it won a hotly-contested British order for 131 aircraft to replace the RAF's ageing Jet Provosts. Considerable modification was undertaken to tailor the basic airframe to exacting requirements, including substituting an 820-kW (1,100-shp) Garrett TPE331-12B turboprop – which significantly improved the rate of climb – and reprofiling the cockpit to provide commonality with the BAe Hawk. EMBRAER flew a Garrett-engined prototype in Brazil in February 1986 and delivered this to Shorts in Belfast as a pattern aircraft. The resultant **Tucano T.Mk 1** retains only 20 per cent commonality with EMBRAER-built aircraft. The first production aircraft flew on 30 December 1986, and initial deliveries to the RAF took place in June 1988. To extend the Tucano's capability in both military training and counter-insurgency roles, Shorts conducted a series of Tucano weapon trials in 1991 with podded machine-guns and LAU-32 rocket launchers. Customers for the armed export Shorts Tucano have been Kenya (12 **T.Mk 51**s) and Kuwait (16 **T.Mk 52**s).

France's 80 EMBRAER-built EMB-312F Tucanos have a ventral airbrake, increased fatigue life, and French avionics.

Shorts Tucano T.Mk 1

Specification: EMBRAER EMB-312 Tucano
Powerplant: one Pratt & Whitney Canada PT6A-25C turboprop rated at 559 kW (750 shp)
Dimensions: wing span 11.14 m (36 ft 6.5 in); length 9.86 m (32 ft 4.25 in); height 3.40 m (11 ft 1.75 in); wing area 19.40 m² (208.82 sq ft)
Weights: basic empty 1810 kg (3,991 lb); normal take-off 2550 kg (5,622 lb); maximum take-off 3175 kg (7,000 lb)
Performance: maximum level speed 'clean' at 3050 m (10,000 ft) 242 kt (448 km/h; 278 mph); maximum cruising speed at 3050 m (10,000 ft) 222 kt (411 km/h; 255 mph); maximum rate of climb at sea level 680 m (2,231 ft) per minute; service ceiling 9145 m (30,000 ft); typical range 995 nm (1844 km; 1,145 miles) with internal fuel; endurance about 5 hours with internal fuel
Armament: four stores stations for maximum ordnance of 1000 kg (2,205 lb), including (7.62-mm (0.30-in) C2 machine-gun pods, 250-kg (551-lb) general purpose bombs; 11-kg (25-lb) practice bombs and rocket launchers

Eurocopter SA 330 Puma

Medium transport/assault helicopter

The AS 532 Cougar is a development of the Puma, featuring a stretched fuselage. This Saudi Arabian AS 532SC carries two Exocet missiles and a search radar for anti-surface vessel duties.

The **Aérospatiale Puma** is the standard medium-lift transport helicopter of the French army and is also in service with air arms around the globe. Pumas can normally carry 15 fully-equipped troops or 2 tonnes of internal cargo (or 2.5 tonnes underslung). Eight prototypes of the **SA 330** were ordered, the first of which flew on 15 April 1965. Initial military production versions comprised: **SA 330B** for the ALAT (Aviation Légère de l'Armée de Terre), **SA 330C** for military export and **SA 330E** (RAF **Puma HC.Mk 1**). Availability of uprated Turmo IVC engines in 1974 better equipped the Puma for 'hot-and-high' operations and the French air force bought 37 of the resulting military **SA 330H** variant as **SA 330Ba**s. Glass-fibre rotors became available in 1977, uprating the H to **SA 330L** standard. The new blades were retrofitted to RAF Pumas and 40 per cent of ALAT's 132 SA 330Bs. Both services' Pumas can carry a pintle-mounted machine-gun in the cabin door, some of the latter's machines additionally receiving nose-mounted radar. One ALAT Puma is testbed for the underfuselage-mounted Orchidée/Horizon battlefield surveillance radar.

Licence-production, exports and export developments

After Aérospatiale ended production with the 686th Puma, the French army bought 15 SA 330Bas and a few attrition replacement helicopters from IAR's Romanian assembly line. The IAR variant can also be equipped with rocket pods, AT-3 Sagger ATMs and fixed machine-gun pods. Isolated by a UN arms embargo, South Africa pursued its own line of Puma development. Two SA 330s were converted as **Atlas XTP-1** gunships which undertook development work for the Rooivalk. In parallel, South Africa has undertaken its own Makila re-engining programme, turning Pumas into **Gemsbok**s. Some civilian-standard Pumas are used by heads of state, while dedicated military operators are Abu Dhabi, Chile, Kuwait, Morocco, Nigeria, Pakistan, Portugal and Zaïre. For the future, in addition to any South African developments, Romania is offering a 'glass cockpit' version as the **Puma 2000**. The Puma is also produced under licence in Indonesia by IPTN.

Among France's wide range of Pumas and Cougars is this AS 332 used for ministerial/presidential transport.

Aérospatiale/Westland (Eurocopter) SA 330 Puma

Specification: Aérospatiale (now Eurocopter France) SA 330L Puma
Powerplant: two Turboméca Turmo IVC turboshafts each rated at 1175 kW (1,575 shp)
Dimensions: main rotor diameter 15.00 m (49 ft 2.5 in); length overall, rotors turning 18.15 m (59 ft 6.5 in) and fuselage 14.06 m (46 ft 1.5 in); height overall 5.14 m (16 ft 10.5 in); main rotor disc area 176.71 m² (1,902.20 sq ft)
Weights: empty 3615 kg (7,970 lb); maximum take-off 7500 kg (16,534 lb); maximum payload 3200 kg (7,055 lb)
Performance: never exceed speed 110 kt (204 km/h; 182 mph); maximum cruising speed 'clean' at optimum altitude 146 kt (271 km/h; 168 mph); maximum rate of climb at sea level 552 m (1,810 ft) per minute; service ceiling 6000 m (19,685 ft); hovering ceiling 4400 m (14,435 ft) in ground effect and 4250 m (13,940 ft) out of ground effect; range 309 nm (572 km; 355 miles)

Eurocopter SA 341/342 Gazelle

Light multi-role helicopter

The SA 342M is the French army's standard anti-armour helicopter, armed with Euromissile HOT weapons. Dedicated anti-helicopter models can carry up to four Mistral AAMs.

Successor to the ubiquitous Sud Alouette II, the **Gazelle** originated in a mid-1960s project by Sud Aviation. Despite using many of its predecessor's dynamic systems (including the 268-kW (360-shp) Astazou II powerplant), the **X.300** design, soon renamed **SA 341**, achieved increased speed and manoeuvrability through adoption of a more powerful turboshaft; aerodynamically-shaped cabin and covered tailboom; and advanced rotor technology. This comprised a rigid main rotor head and glass-fibre blades, and the revolutionary 'fenestron' or fan-in-fin tail rotor. The SA 340 prototype flew on 12 April 1968 with non-standard conventional rotors. The revised **SA 341** incorporated the new rotor technology described above, and introduced a longer cabin, larger tail surfaces and a 440-kW (590-shp) Astazou III.

Six versions were launched initially: **SA 341B** (British Army Gazelle **AH.Mk 1**); **SA 341C** (Royal Navy **HT.Mk 2** trainer); **SA 341D** (RAF **HT.Mk 3** trainer); **SA 341E** (RAF **HCC.Mk 4** VIP transport) – all with Astazou IIIN; **SA 341F** (French army – ALAT) with Astazou IIIC; and military export **SA 341H**. In the UK, Westland built 294 Gazelles, including 212 AH.Mk 1s. Generally unarmed, these carried rockets during the 1982 Falklands War, while nearly 70 were fitted during the late 1980s with target-finding magnifying sights for missile-armed Lynxes. Of 170 SA 341Fs, ALAT converted 40 to carry four HOT ATGMs as **SA 341Ms** and 62 with a 20-mm GIAT cannon and SFOM 80 sight as the **SA 341F/Canon**. Others have acquired an Athos scouting sight.

SA 342

Powered by a 640-kW (858-shp) Astazou XIVH, the **SA 342** replaced the SA 341. Foreign exports began with the military **SA 342K**, the latter soon replaced by **SA 342Ls** with an improved fenestron. The ALAT equivalent is designated **SA 342M** and over 200 have been delivered since 1980, armed with four HOTs and an M397 sight. For the 1991 Gulf War, 30 were converted to **SA 342M/Celtic** with two Mistral AAMs and a SFOM 80 sight. The definitive anti-helicopter model, with four Mistrals and a T2000 sight, will be designated **SA 342M/ATAM** on 30 conversions.

SA 342M Gazelles supported the French army during the Gulf War, taking on Iraqi armour with HOT missiles.

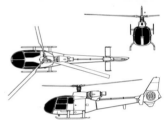

Aérospatiale/Westland (Eurocopter) SA 342 Gazelle

Specification: Aérospatiale (now Eurocopter France) SA 341F Gazelle
Powerplant: one Turboméca Astazou IIIA rated at 440 kW (590 shp)
Dimensions: main rotor diameter 10.50 m (34 ft 5.5 in); fenestron diameter 0.695 m (2 ft 3.375 in); length overall, rotor turning 11.97 m (39 ft 3 in) and fuselage 9.53 m (31 ft 3 in); height overall 3.18 m (10 ft 5.25 in); main rotor disc area 86.59 m² (932.08 sq ft)
Weights: empty 920 kg (2,028 lb); maximum take-off 1800 kg (3,968 lb)
Performance: never exceed speed at sea level 168 kt (310 km/h; 193 mph); maximum cruising speed at sea level 142 kt (264 km/h; 164 mph); maximum rate of climb at sea level 540 m (1,770 ft) per minute; service ceiling 5000 m (16,450 ft); hovering ceiling 2850 m (9,350 ft) in ground effect; 361 nm (670 km; 416 miles)
Armament: maximum payload 700 kg (1,540 lb); SA 341F/Canon carries a GIAT M621 20-mm cannon to starboard and a SFOM 80 sight

Eurocopter SA 365 Dauphin

The AS 565 Panther is the current Dauphin version, now in military use with several nations, including Brazil. The army operates 36 HM-1s, including 10 assembled by Helibras.

In the early 1970s Aérospatiale (now Eurocopter) began development of a helicopter to supersede the Alouette III. The initial version, known as the **SA 360 Dauphin**, featured a four-bladed main rotor, fenestron tail rotor, tailwheel landing gear and standard accommodation for a pilot and up to nine passengers. The 783-kW (1,050-shp) Astazou XVIIIA powered production aircraft, and despite development of a dedicated **SA 361H** military helicopter, it was obvious that the Dauphin's military potential lay with a twin-engined helicopter. The Arriel-powered **SA 365C Dauphin 2** flew for the first time on 24 January 1975. Greater success was achieved by the **SA 365N**, which introduced a retractable tricycle undercarriage, greater use of composites in the construction and other improvements. The Dauphin is built under licence in China as the **Harbin Z-9**.

US Coast Guard Dolphins

Under the company designation **SA 366G1**, Aérospatiale developed a variant of the Dauphin to answer a US Coast Guard requirement to replace its elderly Sikorsky HH-52s. In order to satisfy political requirements, the **HH-65 Dolphin** featured many US-built components, including 507-kW (680-shp) Textron Lycoming LTS101-750A-1 engines. The HH-65 is intended to operate the SRR (short-range recovery) mission from both shore bases and USCG vessels. Mission specific equipment includes inflatable flotation bags, communications and navigation equipment, starboard-side rescue hoist and searchlight and nose-mounted See Hawk FLIR for all-weather rescue operations.

Panther

A multi-role military prototype of the twin-engined **Dauphin** was flown on 29 February 1984 as the **AS 365M**. This can carry 10-12 soldiers, or an armament of eight HOT ATGMs or 44 SNEB rockets. A further developed prototype appeared in April 1986 as the **AS 365K**, for which the name **Panther** was adopted. Varaints now built are the basic armed **AS 565AA**, anti-tank **AS 565CA**, utility **AS 565SA** and unarmed **AS 565MA** for SAR and other naval tasks.

The USCG's HH-65 Dolphins are used for short-range search and rescue work, complementing the larger HH-60 Jayhawks.

Eurocopter (Aérospatiale) HH-65A Dolphin

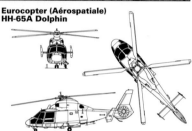

Specification: Eurocopter (Aérospatiale) AS 565UA Panther
Powerplant: two 584-kW (783-shp) Turboméca Arriel IM1 turboshafts each rated at 558 kW (749 shp) for take-off and 487 kW (560 kW) for continuous running
Dimensions: main rotor diameter 11.94 m (39 ft 2 in); fenestron diameter 1.10 m (3 ft 7.5 in); length overall, rotor turning 13.68 m (44 ft 10.625 in) and fuselage 12.11 m (39 ft 8.75 in); height overall 3.99 m (13 ft 1 in); main rotor disc area 111.97 m² (1,205.26 sq ft)
Weights: empty 2193 kg (4,835 lb); normal take-off 4100 kg (9,039 lb); maximum take-off 4250 kg (9,369 lb); maximum payload 1600 kg (3,527 lb)
Performance: maximum cruising speed at sea level 150 kt (278 km/h; 173 mph); maximum rate of climb at sea level 420 m (1,378 ft) per minute; hovering ceiling 2600 m (8,530 ft) in ground effect and 1850 m (6,070 ft) out of ground effect; range 472 nm (875 km; 544 miles) with standard fuel

Eurocopter Tiger/Tigre

This Tigre prototype is seen in Gerfaut escort configuration, equipped with a dummy STRIX roof-mounted sight, a 30-mm nose cannon, a 22-shot 68-mm rocket pod and Mistral AAMs.

The **Eurocopter Tiger** was developed to meet a Franco-Germany requirement for a second-generation anti-tank helicopter (German **PAH-2** and French army **HAC**). Development was begun in 1984, and resumed in March 1987 (after reappraisal) in a modified form to cover a common anti-tank version for the two armies, and an armed escort version (**HAP**) for the French army. A development contract awarded to Eurocopter in November 1989 provides for construction of five prototypes including two in full anti-tank **Tiger** (Germany)/**Tigre** (France) configuration and one as the escort **Gerfaut**.

The Tiger has a slender fuselage with two seats in tandem, stepped and offset to each side of the centreline. The structure makes extensive use of composite materials, and an advanced four-bladed composite semi-rigid main rotor is fitted. Other features include Spheriflex tail rotor and fixed tricycle undercarriage with single wheels and high energy absorption.

In the **PAH-2 Tiger** configuration for the German army, the helicopter was to have been able to carry up to eight HOT 2 or Trigat anti-armour missiles, or four HOTs and four Stinger 2 AAMs for self-defence. Weapons sighting will be by means of a mast-mounted FLIR night-vision system via a pilot's helmet sight. This variant has been abandoned in favour of a utility version, known as **UHU** of as yet unknown configuration but optimised for multi-role duties, including escort work. The Gerfaut will have a roof-mounted TV, FLIR, laser rangefinder and direct-optics sensors.

Prototype flights and planned procurement

The first Tiger prototype flew in April 1991, at first with a mast-mounted sight, but later reconfigured with a canopy sight in the Gerfaut configuration. Estimated requirements (subject to final confirmation) are for 75 HAP and 140 HAC for France and 138 PAH-2/UHU for Germany. It is expected that the French army will receive Gerfauts in 1994 and Tigres in 1999, with UHU Tigers reaching the Bundeswehr in 1998.

The Tigre will replace the Gazelle in the anti-tank role with the French army. British Aerospace is teamed with Eurocopter to offer the type to the British army.

Eurocopter HAC Tiger

Specification: Eurocopter HAC Tigre/PAH-2 Tiger
Powerplant: two MTU/Turboméca/Rolls-Royce MTR 390 turboshafts each rated at 958 kW (1,285 shp) for take-off and 873 kW (1,171 shp) for continuous running
Dimensions: main rotor diameter 13.00 m (42 ft 7.75 in); fuselage legth 14.00 m (45 ft 11.25 in); height overall 4.32 m (14 ft 2 in) to top of turning tail rotor; main rotor disc area 132.73 m² (1,428.76 sq ft)
Weights: basic empty 3300 kg (7,275 lb); normal take-off 5800 kg (12,787 lb); maximum overload take-off 6000 kg (13,227 lb)
Performance: maximum cruising speed at optimum altitude 151 kt (280 km/h; 174 mph); maximum rate of climb at sea level more than 600 m (1,969 ft) per minute; endurance over 3 hours
Armament: (HAC Tigre) primary armament of MATRA Mistral AAMs for defence; (HAP Gerfaut) one GIAT AM-30781 30-mm cannon in undernose turret and two 22-round 68-mm unguided SNEB rockets plus either four Mistral AAMs or two 12-round rocket pods on stub wing pylons

Eurofighter EFA 2000

EFA is scheduled to become the standard multi-role fighter for Germany, Italy, Spain and the UK. This is the first prototype (DA1), assembled by DASA in Germany.

The **Eurofighter** consortium was formed in June 1986 by Britain, Germany and Italy (soon joined by Spain) to produce an air superiority fighter by the late 1990s. Much experience was gained with main **EFA** (European Fighter Aircraft) concepts from BAe's Experimental Aircraft Programme (EAP), including the unstable canard delta layout, active digital fly-by-wire flight controls, advanced avionics, multi-function cockpit displays, carbon-fibre composite construction and extensive use of aluminium-lithium alloys and titanium. The twin-RB.199 EAP technology demonstrator first flew in August 1986 and amassed invaluable data before retirement in May 1991. A development contract signed in late 1988 covered building and testing until 1999 of eight EFA prototypes (now reduced to seven). Eventual purchase was envisaged from 1996 of 765 EFAs, comprising 250 each for the RAF and the Luftwaffe, 165 for the AMI and 100 for the Ejercito del Aire.

ECR-90 radar and defensive systems

After major contention, in May 1990 the new ECR-90 multimode pulse-Doppler look-up/look-down radar was selected for development over an uprated Hughes APG-65. While optimised for AMRAAM use, ECR-90 also provides CW illumination for SARH AAMs. The radar is supplemented by an IRST. Integrated defensive aids comprise missile approach, laser and radar warning systems, wingtip ESM/ECM pods, chaff/flare dispensers and a towed decoy.

Project reappraisal and first flight

In 1992, cost studies challenged system unit price estimates as unaffordable, and resulted in major project reviews against threats of an FRG withdrawal. After reappraisal, a slightly simplified and less capable **New EFA** or **Eurofighter 2000** variant was agreed upon in late 1992, with options for cheaper individual equipment fits. The total procurement has now been reduced to less than 600 aircraft, comprising UK (250), Germany (138), Italy (130) and Spain (72). The prototype EFA 2000 made its long-awaited first flight on 27 March 1994 from Manching, Germany.

The sole BAe EAP technology demonstrator conducted a valuable programme and proved much of the EFA concept.

Eurofighter EFA 2000

Specification: Eurofighter EFA
Powerplant: two Eurojet EJ200 turbofans each rated at about 60.0 kN (13,490 lb st) dry and 90.0 kN (20,250 lb st) with afterburning
Dimensions: wing span 10.50 m (34 ft 5.5 in); length 14.50 m (47 ft 7 in); height about 4.00 m (13 ft 1.5 in); wing area 50.00 m² (538.21 sq ft); canard foreplane area 2.40 m² (25.83 sq ft)
Weights: empty 9750 kg (21,495 lb); maximum take-off 21000 kg (46,297 lb)
Performance: maximum level speed 'clean' at 11000 m (36,090 ft) 1,147 kt (2125 km/h; 1,321 mph) take-off run 500 m (1,640 ft) at normal take-off weight; landing run 500 m (1,640 ft) at normal landing weight; combat radius between 250 and 300 nm (463 and 556 km; 288 and 345 miles)
Armament: one internal Mauser Mk 27 27-mm cannon in the starboard fuselage, four BVR AAMs carried semi-recessed in low-drag fuselage stations, with nine other stores pylons (three for drop tanks), having a 6500-kg (14,330-lb) total capacity

49

Fairchild A-10 Thunderbolt II

Zap! The A-10's 30-mm cannon is lethal against armour. Although attempts were made to divert gun gases away from the aircraft, the plain barrel end has been retained.

Fairchild A-10A Thunderbolt II

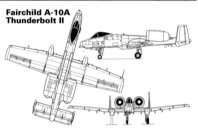

Originally conceived as a counter-insurgency aircraft for the Vietnam War, the **Fairchild A-10A Thunderbolt II** emerged in 1972 as a dedicated close air support aircraft, with a primary anti-armour role. The A-10's operating environment dictated a highly-survivable design incorporating a large-area wing for excellent low-altitude manoeuvrability, rear-mounted engines shrouded from ground fire by either the wings or tailplane and redundant, armoured and duplicated flight controls and hydraulic systems. Titanium 'bathtubs' protect both the pilot and the ammunition tank.

Missile and cannon armament

The principal weapon is the AGM-65 Maverick anti-armour missile, supplemented by an enormous GAU-8/A 30-mm seven-barrelled rotary cannon. Avionics of the A-10 remained very basic for most of the aircraft's career, with no laser designator or rangefinder fitted. The pilot has a HUD, and a screen for displaying images from Maverick or other EO-guided weapons. A Pave Penny seeker spots targets designated by laser. Most current aircraft have received the LASTE modification, which finally adds an autopilot and also improves gun accuracy.

Squadron service and combat

The A-10 entered USAF service in 1977. At its peak deployment, a total of six European A-10 squadrons was stationed in England. Debates raged as to the vulnerability of the A-10, and it was finally decided to gradually withdraw the type in favour of the F-16. At the same time, redundant A-10s became available to replace OV-10s in the forward air control role. Unchanged, these were redesignated **OA-10A** and distributed to tactical air support squadrons. For the FAC role the A-10s are armed with AIM-9 AAMs for self-defence and rocket pods for marking targets. In the twilight of its career, the A-10 proved its worth during 1991 Desert Storm operations, destroying huge numbers of tanks, artillery pieces and vehicles. The A-10 was not exported, although it seems likely that some 50 airframes may be sold to Turkey.

Some USAF A-10s have acquired a forward air control role, for which they are armed with rocket pods and AIM-9 missiles.

Specification: Fairchild Republic OA-10A Thunderbolt II
Powerplant: two General Electric TF34-GE-100 turbofans each rated at 40.32 kN (9,065 lb st)
Dimensions: wing span 17.53 m (57 ft 6 in); 16.26 m (length 53 ft 4 in); height 4.47 m (14 ft 8 in); wing area 47.01 m² (506.00 sq ft)
Weights: basic empty 9771 kg (21,541 lb); operating empty 11321 kg (24,959 lb); forward airstrip armed 14865 kg (32,771 lb); maximum take-off 22680 kg (50,000 lb)
Performance: maximum level speed 'clean' at sea level 381 kt (706 km/h; 439 mph); maximum rate of climb at sea level 1828 m (6,000 ft) per minute; combat radius 540 nm (1000 km; 620 miles) on a deep strike mission or 250 nm (463 km; 288 miles) on a close air support mission with a 1.7-hour loiter
Armament: one internal GAU-8A 30-mm cannon with 1,174 rounds; twin AIM-9L installation and one ALQ-184 or ALQ-131 ECM pod for self-defence, and up to 12 LAU-68 rocket pods for marking targets; maximum theoretical ordnance of 7258 kg (16,000 lb)

General Dynamics F-111

The F-111F is the last major USAF variant in service. It is highly effective as a launch platform for laser-guided bombs using the Pave Tack acquisition system.

Developed to meet a joint service requirement for a long-range interceptor and deep-strike interdictor, the controversial **General Dynamics F-111** first flew in December 1964. Many notable innovations were first introduced, including the variable-geometry wing, afterburning TF30 turbofans and a cockpit escape capsule. The Navy's **F-111B** interceptor was cancelled due to severe weight problems. Most early 'Aardvark' models are no longer in service, including the original **F-111A** production variant and **F-111D** with uprated engines and a radically updated 'digital' avionics system. Although highly capable, The D proved to be maintenance intensive and was retired in late 1992. Some of SAC's **FB-111A** strategic bombers with longer-span wings were reworked as **F-111G**s, and these served in a training role until 1993. The only early-generation aircraft currently left in USAF service are 24 AMP-avionics ugraded **F-111E**s.

Export sale – Australian F-111Cs
From 1973, the RAAF received 24 long-span **F-111C**s, later augmented by four F-111As and 24 surplus FB-111As. RAAF F-111Cs can be equipped with the Pave Tack acquisition/designation pod, laser-guided bombs, GBU-15 EO-guided bombs, indigenous Karinga cluster bombs, AGM-84 Harpoon anti-ship missiles and AGM-88 HARMs. Four **RF-111C**s were modified from F-111Cs with an internally-mounted multi-sensor reconnaissance pallet.

F model
Last of the production variants, the **F-111F** is also the last in operational service with the US Air Force. It introduced less sophisticated, but more reliable, avionics, and significantly uprated engines. The F-111F's weapons bay is now used primarily to carry a Pave Tack pod, which incorporates a FLIR sensor and bore-sighted laser rangefinder/designator for autonomous LGB operations, a capability which was graphically demonstrated during Desert Storm. The 84 surviving aircraft are undergoing the Pacer Strike update programme, replacing analog systems with digital avionics.

Australia was the only export customer for the F-111. Its fleet includes four RF-111C reconnaissance platforms.

General Dynamics F-111F

Specification: General Dynamics F-111F
Powerplant: two Pratt & Whitney TF30-P-100 turbofans each rated at 111.65 kN (25,100 lb st) with afterburning
Dimensions: wing span 19.20 m (63 ft 0 in) spread and 9.74 m (31 ft 11.4 in) swept; length 22.40 m (73 ft 6 in); height 5.22 m (17 ft 1.4 in); wing area 48.77 m² (525.00 sq ft) spread
Weights: operating empty 21537 kg (47,481 lb); maximum take-off 45360 kg (100,000 lb)
Performance: maximum level speed 'clean' at 10975 m (36,000 ft) 1,433 kt (2655 km/h; 1,650 mph); cruising speed at high altitude 496 kt (919 km/h; 571 mph); range more than 2,540 nm (4707 km; 2,925 miles) with internal fuel
Armament: two AIM-9P-3 AAMs for self-defence; maximum ordnance of 14228 kg (31,500 lb), primary weapons are 500-lb GBU-12, 2,000-lb GBU-10, 2,000-lb GBU-24 LGBs, 4,800-lb GBU-28 'Deep Throat' and GBU-15 2,000-lb EO-guided bombs; 'dumb' ordnance includes iron bombs, cluster weapons, BLU-107 Durandal anti-runway and B61 nuclear weapons

51

Grumman A-6 Intruder

Two-seat carrier-based attack aircraft

The Intruder is the main strike asset available to the carrier battle group. It can carry a wide range of stores, including laser-guided bombs, anti-ship and anti-radar missiles.

In December 1957 the **Grumman A-6 Intruder** was selected to fulfil a US Navy requirement for a new long-range, low-level tactical strike aircraft. Eight development **A-6A**s were ordered and the first flight was made on 19 April 1960. The first of 482 production A-6As was delivered to the USN in February 1963, with Vietnam combat operations commencing in March 1965. The Intruder's Digital Integrated Attack Navigation Equipment (DIANE) endowed an excellent operating ability in all weather conditions. A-6A procurement included 21 USMC **EA-6A**s for ECM support and Elint gatherering. Converted A-6A variants comprised: 19 **A-6B**s for SAM suppression; 12 **A-6C**s with turret-mounted FLIR and LLLTV equipment for improved night-attack capability and 78 **KA-6D** inflight-refuelling tankers, equipped with TACAN and mounting a hose-reel unit in the rear fuselage to refuel other aircraft under the 'buddy' system. The KA-6D is the USN's standard carrier-based tanker.

Improved A-6E and A-6E developments

The upgraded **A-6E** was first flown on 27 February 1970 and introduced a multi-mode navigation/attack radar and a computerised nav/attack system. Nine A-6Es were additionally converted as KA-6Ds. All 445 USN and USMC A-6Es were progressively updated further from 1974 with a TRAM laser designator and FLIR turret giving 'smart' weapon capability. The **A-6F** is a revised A-6E airframe with new radar, digital avionics and F404 turbofans. This was cancelled when the Navy pursued the A-12 programme, but Grumman continued development of the **A-6G**, basically the F model but retaining the ageing J52 engines. After the A-12's cancellation, the A-6E fleet has been the subject of continuing upgrades. Rewinging with composite units was considered a necessity to keep the fleet airworthy, while the **SWIP** (Systems/Weapons Integration Programme) introduced increased weapons capability. Further updating was implemented in late 1992 with the first flight of the Block 1-A upgrade machine, introducing pilot's HUD, revised wing fillets and extra fuel.

The first SWIP A-6s entered service in 1990/91, and can carry stand-off weapons such as AGM-65 Maverick, AGM-84 Harpoon Block 1C, AGM-84E SLAM and AGM-88 HARM.

Grumman A-6E Intruder

Specification: Grumman A-6E Intruder SWIP
Powerplant: two Pratt & Whitney J52-P-8B turbojets each rated at 41.37 kN (9,300 lb st)
Dimensions: wing span 16.15 m (53 ft 0 in); width folded 7.72 m (25 ft 4 in); length 16.69 m (54 ft 9 in); height 4.93 m (16 ft 2 in); wing area 49.13 m² (528.90 sq ft)
Weights: empty 12525 kg (27,613 lb); maximum take-off 26580 kg (58,600 lb) for catapult launch or 27397 kg (60,400 lb) for field take-off
Performance: maximum level speed 'clean' at sea level 560 kt (1037 km/h; 644 mph); cruising speed at optimum altitude 412 kt (763 km/h; 474 mph); maximum rate of climb at sea level 2323 m (7,620 ft) per minute; service ceiling 12925 m (42,400 ft); range with maximum warload 878 nm (1627 km; 1,011 miles)
Armament: maximum ordnance 8165 kg (18,000 lb) including AGM-65 Maverick, AGM-84 Harpoon, AGM-84E SLAM and AGM-88 HARM missiles, typical bomb loads of up to 30 227-kg (500-lb) Mk 82, or three 907-lb (2,000-lb) Mk 84 bombs

Grumman EA-6B Prowler

The EA-6B carries the main group of radar receivers in the fin-top 'football' fairing. The ALQ-99F jamming pod is carried underwing and is powered by an external ram air turbine.

The **EA-6B Prowler** is the USN's standard carrierborne electronic warfare aircraft. Early experience with the **EA-6A** EW platform led to the development of an advanced, lengthened four-seat A-6 variant, seating a pilot and three electronic warfare officers (EWOs) to manage the sophisticated array of ECM and ESM systems. Entering service during 1971, it introduced a tactical jamming system (TJS) which employs 'noise' jamming originating from a maximum of five transmitter pods. The first 23 production aircraft were to 'Basic' standard with ALQ-99 TJS and ALQ-92 comms jamming system.

Grumman EA-6B Prowler

Improved EXCAP and ICAP models

These were followed in 1973 by 25 **EXCAP** (Expanded Capability) airframes with ALQ-99A TJS. In 1976, the **ICAP** (Improved Capability) standard was applied to 45 new-build and 17 previous machines and introduced new displays, AN/ALQ-126 multiple-band defensive breakers and updated radar deception gear. All 55 surviving ICAPs were upgraded with software and display improvements to **ICAP-II** standard. This also introduces upgraded jamming pods and is able to handle groups of weapons systems with improved identification of hostile emitters and improved reliability and maintainability. The **ICAP-II/Block 86** introduces 'hard kill' capability with AGM-88A HARMs.

ADVCAP and aerodynamic improvements

Most recently, EA-6Bs have been upgraded to two **ADVCAP** configurations. The basic ADVCAP has new jammer transmission and passive detection capabilities and an expanded AN/ALE-39 chaff dispenser fit. An Avionics Improvement Program will lead to a remanufactured **ADVCAP/Block 91** EA-6B with new displays, radar improvements, an improved tactical support jamming suite, AN/ALQ-149 communications jamming system and a digital autopilot. Aerodynamic improvements were developed under the **VEP** (Vehicle Improvement Program) project and comprise the addition of fuselage strakes, modified flaps, slats, speed brakes and a fin extension.

Launch! The Prowler is vital to protecting the carrier and its air wing during offensive operations.

Specification: Grumman EA-6B Prowler
Powerplant: two Pratt & Whitney J52-P-408 turbojets each rated at 49.8 kN (11,200 lb st)
Dimensions: wing span 16.15 m (53 ft 0 in); width folded 7.87 m (25 ft 10 in); length 18.24 m (59 ft 10 in); height 4.95 m (16 ft 3 in); wing area 49.13 m² (528.90 sq ft)
Weights: empty 14321 kg (31,572 lb); normal take-off from a carrier in stand-off jamming configuration with five jammer pods 24703 kg (54,461 lb) or from land with maximum internal and external fuel 27493 kg (60,610 lb); maximum take-off 29484 kg (65,000 lb)
Performance: maximum level speed with five jammer pods 530 kt (982 km/h; 610 mph); cruising speed at optimum altitude 418 kt (774 km/h; 481 mph); maximum rate of climb with five jammer pods 3057 m (10,030 ft) per minute; service ceiling with five jammer pods 11580 m (38,000 ft); range 955 nm (1769 km; 1,099 miles) with maximum external load
Armament: (ADVCAP/Block 91) up to four AGM-88A HARM anti-radar missiles

Grumman E-2 Hawkeye

Japan's 13 E-2C Hawkeyes are based at Misawa, on the north of Honshu island. From here, early warning patrols are mounted over the waters of northern Japan.

The **E-2 Hawkeye** has been the US Navy's airborne early warning platform since entering service in 1964. It has a rotodome (mounted above the rear upper fuselage) housing antennas for the main radar and IFF systems. Including prototypes and development aircraft, a total of 59 **E-2As** was built, featuring APS-96 surveillance radar. Most were later converted to **E-2B** standard with a general-purpose computer and retired from service in the mid-1980s.

Current generation Hawkeyes are to **E-2C** standard, the first example of which flew on 20 January 1971. Identified by a cooling intake behind the cockpit, the E-2C introduced a new APS-125 radar and much improved signal processing capability. The basic E-2C has been the subject of continual updating over the years. Radar units have changed to APS-138, APS-139 (from 1989) and APS-145, currently fitted to new aircraft and retrofitted to earlier aircraft. APS-145 offers improved resistance to jamming and better overland surveillance capability. Other potential upgrades concern the IFF system and installation of JTIDS. Production is almost at an end with the delivery of the 139th aircraft for the US Navy.

Grumman E-2C Hawkeye

Operational profile

In the AEW role, the E-2C extends the detection range of the battle group by about 480 km (300 miles) for aircraft and 258 km (160 miles) for cruise missiles. Surface vessels can also be located. Constant communication is maintained with the carrier's Combat Information Center and patrolling F-14 fighters by means of a datalink. The Hawkeye can also act as an airborne control and command post, supplying directions to attack aircraft and escorting fighters to deconflict them, in addition to providing warnings of hostile aircraft.

In USN service, the E-2C flies with 12 active-duty units, two training units and two Reserve squadrons. E-2Cs have been exported to Egypt (five plus one on order), Japan (13), Singapore (four plus two on option) and Taiwan (four). Thailand and France each have requirements for four E-2s.

Commonly known as the 'Hummer', the E-2 flies with every US Navy carrier air wing. It is used to control aircraft during strikes and for airborne early warning for the battle group.

Specification: Grumman E-2C Hawkeye
Powerplant: two Allison T56-A-425 turboprops each rated at 3661 kW (4,910 ehp)
Dimensions: wing span 24.56 m (80 ft 7 in); folded width 8.94 m (29 ft 4 in); length 17.54 m (57 ft 6.75 in); height 5.58 m (18 ft 3.75 in); wing area 65.03 m² (700.00 sq ft)
Weights: empty 17265 kg (38,063 lb); maximum take-off 23556 kg (51,933 lb)
Performance: maximum level speed 323 kt (598 km/h; 372 mph); maximum cruising speed at optimum altitude 311 kt (576 km/h; 358 mph); ferry cruising speed at optimum altitude 268 kt (496 km/h; 308 mph); maximum rate of climb at sea level 767 m (2,515 ft) per minute; service ceiling 9390 m (30,800 ft); minimum take-off run 610 m (2,000 ft); take-off distance to 15 m (50 ft) 793 m (2,600 ft) at maximum take-off weight; minimum landing run 439 m (1,440 ft); 1,605 miles); operational radius 175 nm (320 km; 200 miles) for a patrol of 3 to 4 hours

Grumman F-14A Tomcat

F-14As of VF-84 'Jolly Rogers' enter the carrier approach pattern with arrester hooks deployed. Variable-geometry wings enable landing speeds to be kept acceptably low.

Designed as a successor to the F-4 in the fleet air defence role, the **Grumman F-14A Tomcat** was originally conceived to engage and destroy targets at extreme range. It remains the US Navy's standard carrier-based interceptor. A dozen development aircraft were ordered, with the first making its maiden flight on 21 December 1970.

Advanced radar and weapons

The key to the F-14's effectiveness lies in its advanced avionics suite, the AWG-9 fire control system representing the most capable long-range interceptor radar in service, with the ability to detect, track and engage targets at ranges in excess of 160 km (100 miles). The early IRST system was later replaced (and retrofittted) with the TCS long-range camera. Primary armament comprises Hughes' AIM-54 Phoenix (which remains the longest-ranged air-to-air missile in service today), AIM-7 or AIM-120 AMRAAM for medium range, and AIM-9 for short-range, close-in engagements, augmented by the fixed internal 20-mm cannon.

Service

Deliveries to the Navy began in October 1972, with the first operational cruise in 1974. Production continued into the 1980s and a total of 26 front-line and four second-line squadrons were eventually equipped with the F-14A. Since entering fleet service, the F-14 has suffered many difficulties, many related to the troublesome TF30 turbofan: fan blade losses and compressor stall-related problems caused several losses. The problems were largely solved with the revised TF30-P-414A version of the powerplant.

Additional roles

F-14As are also used for reconnaissance missions, carrying the Tactical Air Reconnaissance Pod System (TARPS), and it is usual for three TARPS-capable aircraft to be assigned to one of each carrier air wing's F-14 units. More recently, the F-14A has also acquired a secondary air-to-ground role. The 'Bombcat' carries only conventional 'iron' bombs, and has no autonmous PGM capability.

Naval Air Reserve squadron VF-202 acquired F-14As in 1987 and is the TARPS unit for its reserve air wing (CVWR-20).

Grumman F-14A Tomcat

Specification: Grumman F-14A Tomcat
Powerplant: two Pratt & Whitney TF30-P-412A/414A turbofans each rated at 92.97 kN (20,900 lb st) with afterburning
Dimensions: wing span 19.54 m (64 ft 1.5 in) spread, 11.65 m (38 ft 2.5 in) swept; length 19.10 m (62 ft 8 in); height 16 ft 0 in (4.88 m); wing area 52.49 m² (565.00 sq ft)
Weights: empty 18191 kg (40,104 lb) with -414A engines; normal take-off 'clean' 26632 kg (58,715 lb); maximum take-off 32098 kg (70,764 lb) with six AIM-54s; overload take-off 33724 kg (74,349 lb)
Performance: maximum level speed 'clean' at high altitude 1,342 kt (2485 km/h; 1,544 mph) and at low altitude 792 kt (1468 km/h; 912 mph); maximum rate of climb at sea level more than 9145 m (30,000 ft) per minute; service ceiling more than 15240 m (50,000 ft); radius on a CAP 665 nm (1,233 km; 766 miles)
Armament: one internal M61 Vulcan 20-mm cannon in lower forward fuselage with 675 rounds; (typical intercept) two AIM-54C Phoenix, two AIM-7 Sparrow and two AIM-9 Sidewinder AAMs

Grumman F-14B/D Tomcat

Larger and more smoothly-contoured jetpipes identify this Tomcat as an F-14B. VF-101 is a fleet replenishment squadron and trains F-14 pilots and radar intercept officers.

Problems with the F-14A's TF-30 turbofan were a key factor in the development of re-engined and upgraded Tomcat variants. One of the original prototype airframes was fitted with two F401-PW-400s and employed for an abbreviated test programme as the **F-14B** as early as 1973-74. Technical problems and financial difficulties forced the abandonment of the programme, and the aircraft was placed into storage, re-emerging as the **F-14B Super Tomcat** with F101DFE engines. This engine was developed into the GE F100-GE-400 turbofan, which was selected to power production improved Tomcat variants. Two re-engined variants were proposed: the **F-14A+** was to be an interim type (primarily concerning F-14A conversions), while the **F-14D** would also feature improved digital avionics. Subsequently, the F-14A(Plus) was formally redesignated as the F-14B, 38 new-build examples being joined by 32 F-14A rebuilds in equipping six deployable squadrons from 1988. These incorporated some avionics changes, including a modernised fire control system, new radios, upgraded RWRs, and various cockpit changes.

F-14D

Two modified F-14As served as F-14D prototypes and the first D model to be built as such made its maiden flight on 9 February 1990. The F-14D added digital avionics, with digital radar processing and displays (adding these to standard AWG-9 hardware under the redesignation APG-71), and a dual undernose TCS/IRST sensor pod. Other improvements introduced by the F-14D include NACES ejection seats, and AN/ALR-67 radar warning receiver equipment. The F-14D also has expanded ground attack capability. The US Department of Defense's decision to cease F-14D funding has effectively halted the Navy's drive to upgrade its force of Tomcats. In consequence, the service has received only 37 new-build examples of the F-14D, with the first deliveries effected in November 1990. Plans to upgrade approximately 400 existing F-14As to a similar standard have also been severely curtailed and only about 18 have been updated.

The F-14B and D have significantly more thrust than the F-14A, redressing the earlier variant's power shortage.

Grumman F-14B Tomcat

Specification: Grumman F-14D Tomcat
Powerplant: two General Electric F110-GE-400 turbofans each rated at 62.27 kN (14,000 lb st) dry and 102.75 kN (23,100 lb st) with afterburning
Dimensions: wing span 19.54 m (64 ft 1.5 in) spread, 11.65 m (38 ft 2.5 in) swept and 10.15 m (33 ft 3.5 in) overswept; length 19.10 m (62 ft 8 in); height 4.88 m (16 ft 0 in); wing area 52.49 m² (565.00 sq ft)
Weights: empty 18951 kg (41,780 lb); normal take-off 29072 kg (64,093 lb) for a fighter/escort mission or 33157 kg (73,098 lb) on a fleet air defence mission; maximum take-off 33724 kg (74,349 lb)
Performance: maximum level speed 'clean' at high altitude 1,078 kt (1997 km/h; 1,241 mph); cruising speed at optimum altitude 413 kt (764 km/h; 475 mph); maximum rate of climb at sea level more than 9145 m (30,000 ft) per minute; service ceiling more than 16150 m (53,000 ft); combat radius on a combat air patrol with six AIM-7s and four AIM-9s 1,075 nm (1994 km; 1,239 miles)
Armament: one 20-mm cannon; (typical intercept) two AIM-54C Phoenix, two AIM-7 Sparrow and two AIM-9 Sidewinder AAMs

Grumman/GD EF-111 Raven

The EF-111A performs three primary tasks: provision of a stand-off barrage to disguise raids, direct escort of strike aircraft deep into enemy territory, and battlefield jamming support.

Based on the original F-111A production variant of the General Dynamics swing-wing strike fighter, the **EF-111A Raven** evolved as specialised electronic warfare platform capable of undertaking stand-off and penetration escort missions. Development of the EF-111A was initiated in 1972, and Grumman first flew two converted prototypes in 1977. The tactical jamming system that forms the core of the EF-111A's impressive capability is a variation of the AN/ALQ-99 TJS as fitted to Grumman's EA-6B. The Raven introduces a much improved equipment package. It also embodies a greater degree of automation so as to allow it to be managed by just one EWO. Receiving antennas associated with the TJS are located in a distinctive bulbous fin-cap fairing, while the jamming transmitters are contained in space previously occupied by the internal weapons bay. These are housed in a 5-m (16-ft) long ventral 'canoe' radome. Other EW equipment include active and passive ECM systems and an IR warning system. The EF-111A has no armament capability and is therefore forced to rely on high-speed evasion in the event of running into fighters.

Procurement and upgrade

A total of 42 aircraft was brought to EF-111A standard by Grumman and most of these are still operational. Survivors were earmarked for an extensive Service Improvement Program (SIP) intended to upgrade existing capability and enhance reliability to counter progressively more sophisticated radar threats. SIP may be reduced for funding reasons. However, the EF-111A remains the only dedicated jamming platform in the USAF inventory and is likely to remain unique in that capacity for the foreseeable future.

Deployment

The EF-111A was committed to combat during Operation Desert Storm when it operated from bases at Taif, Saudi Arabia, and Incirlik, Turkey. With the withdrawal in 1992 of the EF-111A squadron from Upper Heyford, England, the Raven force is now concentrated in the US under the control of ACC with the 27th Fighter Wing at Cannon AFB, NM.

The large degree of automation of the EW system of the 'Spark Vark' enables just one EWO to perform this task.

Grumman/General
Dynamics
EF-111A Raven

Specification: Grumman/General Dynamics EF-111A Raven
Powerplant: two Pratt & Whitney TF30-P-3 turbofans each rated at 82.29 kN (18,500 lb st) with afterburning
Dimensions: wing span 19.20 m (63 ft 0 in) spread and 9.74 m (31 ft 11.4 in) swept; length 23.16 m (76 ft 0 in); height 6.10 m (20 ft 0 in); wing area 48.77 m² (525.00 sq ft) swept and 61.07 m² (657.07 sq ft) spread
Weights: operating empty 25072 kg (55,275 lb); normal take-off 31752 kg (70,000 lb); maximum take-off 40347 kg (88,948 lb)
Performance: maximum speed at high altitude 1,226 kt (2272 km/h; 1,412 mph); maximum combat speed 1,196 kt (2216 km/h; 1,377 mph); average speed in combat area 507 kt (940 km/h; 584 mph); maximum rate of climb at sea level 1006 m (3,300 ft) per minute; service ceiling 13715 m (45,000 ft); combat radius 807 nm (1495 km; 929 miles); unrefuelled endurance more than 4 hours

IAI (Israeli Aircraft Industries) Kfir

Multi-role fighter

Two-seat Kfirs feature a lengthened and drooped nose for improved view. This houses the avionics displaced from the C2's spine and is fitted with small vortex-generating strakes.

The development of the **IAI Kfir** was made possible by Israel's purchase of the F-4 Phantom, and especially its GE J79 engine. The first J79-engined Mirage was a IIIBJ trainer, and this first flew on 19 October 1970. It was joined by a re-engined Nesher (an unlicenced Mirage 5 copy) in 1971. The J79's greater mass flow and higher operating temperature necessitated the provision of enlarged air intakes, a dorsal tailfin air scoop for afterburner cooling and extensive heat shielding of the rear fuselage.

The basic Kfir was produced in small numbers (27) and most were later upgraded as **Kfir C1**s, with narrow-span canards on the intakes and rectangular strakes behind the ranging radar, on the sides of the nose. Twenty-five survivors were leased to the US Navy and USMC in the late 1980s for adversary training under the designation **F-21A**.

Kfir C2

Revealed in 1976, the **Kfir C2** was the first full-standard variant, equipped from the outset with nose strakes, large fixed canard foreplanes and a dogtooth wing leading edge. These aerodynamic features improved turn and take-off performance and improved controllability. New avionics were introduced, including a ranging radar, twin-computer flight control system, multi-mode navigation and weapons delivery system, central air data computer and HUD. IAI built 185 C2s and **TC2** trainers, and about 120 remain in service. After long delays in gaining US approval to re-export the J79 engine, C2s were sold to Ecuador (12) and Colombia (11). Ecuador's C2s appear to have been upgraded to C7 standard. Each country also operates two TC2s.

The **Kfir C7/TC7** designation applies to upgraded aircraft delivered from 1983 onwards. These introduced a HOTAS cockpit, various avionics and cockpit display improvements, and 'smart' weapon capability. Its advanced RWR equipment was also fitted to late C2s. Other changes were an engine overspeed provision ('combat plus'), addition of two further stores stations below the intake ducts, and provision for an optional IFR probe or refuelling receptacle.

The Kfir C7 is used primarily as a fighter-bomber in Israeli service. It has two extra hardpoints below the intake trunks, as seen in this view of a fully-armed IDF/AF aircraft.

IAI Kfir C7

Specification: IAI Kfir-C7
Powerplant: one IAI Bedek Division-built General Electric J79-J1E turbojet rated at 52.89 kN (11,890 lb st) dry and 83.40 kN (18,750 lb st) with afterburning and 'combat boost' option
Dimensions: wing span 8.22 m (26 ft 11.6 in); canard foreplane span 3.73 m (12 ft 3 in); length 15.65 m (51 ft 4.25 in) including probe; height 4.55 m (14 ft 11.25 in); wing area 34.80 m² (374.60 sq ft); canard foreplane area 1.66 m² (17.87 sq ft)
Weights: empty about 7285 kg (16,060 lb); normal take-off 10415 kg (22,961 lb); maximum take-off 16500 kg (36,376 lb)
Performance: maximum level speed 'clean' at 36,000 ft (10975 m) more than 1,317 kt (2440 km/h; 1,516 mph) or at sea level 750 kt (1389 km/h; 863 mph); maximum rate of climb at sea level 14000 m (45,930 ft) per minute; zoom climb ceiling 22860 m (75,000 ft); combat radius 476 nm (882 km; 548 miles) on a 1-hour CAP
Armament: two Rafael-built DEFA 553 30-mm cannon with 140 rpg, maximum ordnance of 6085 kg (13,415 lb), including Shafrir 2 and Python 3 IR AAMs; Mk 80 series bombs; Shrike, Maverick and GBU-13 guided weapons, TAL-1 and -2 CBUs

Ilyushin Il-18/-20/-22/-38

The modifed Il-38 'May' is the Russian navy's standard ASW patroller. Mission equipment includes a search radar (housed in a large ventral radome) and a tail-mounted 'MAD' sting.

Designed primarily as a turboprop airliner for Aeroflot, the **Il-18 'Coot'**, which first flew in July 1957, remains in limited service with a variety of civil and military operators worldwide for transport and VIP transport duties. Production versions accommodated 75-122 passengers and adopted Ivchyenko AI-20 engines as standard. In Russia, a large number of nominally civilian Il-18s serve as equipment and avionics testbeds, and in the experimental role. The replacement of the Il-18 resulted in a pool of redundant airframes suitable for conversion to other roles.

The first of these was the **Il-20 'Coot-A'**, which is presumed to be a dedicated Elint and radar reconnaissance aircraft. It carries a large, cylindrical underfuselage pod (assumed to house a SLAR), smaller forward fuselage pods for cameras or other optical sensors and various antennas on the fuselage and wingtips. The **Il-22 'Coot-B'** is believed to be an airborne command post or communications relay variant. It features a cylindrical pod on the tailfin, and an array of blade antennas above and below the fuselage.

Sub hunter – Il-38 'May'

Like the contemporary Lockheed Electra, the Il-18 airliner formed the basis for a long-range ASW aircraft. However, the extent of the changes to the resulting **Il-38 'May'** make it virtually certain that it is a new-build aircraft, and not a conversion. The basic Il-18 fuselage has been lengthened by approximately 4 m (13 ft 1.5 in), and the wings have been moved forward to compensate for the effect on aircraft centre of gravity of the new role equipment. Structural alterations include the provision of a MAD stinger projecting aft from the tailcone, and a pair of internal weapons/stores bays fore and aft of the wing structure. The Il-38 has a weather radar in the nose, with its search radar housed in a distinctive, bulged radome below the forward fuselage. The external skin is disrupted by a number of antennas, heat exchanger outlets, and by large heat exchanger inlet pods and cable ducts just ahead of the wing. The sole export customer for the Il-38 is the Indian navy, which received five 'Mays'.

Russia's Il-20 'Coot-A' electronic intelligence gatherer is identified by its massive ventral 'canoe' radome.

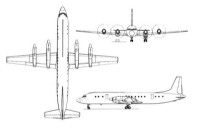

Ilyushin Il-20 'Coot-A'

Specification: Ilyushin Il-38 'May'
Powerplant: four ZMDB Progress (Ivchyenko) AI-20M turboprops each rated at 3169 kW (4,250 shp)
Dimensions: wing span 37.42 m (122 ft 9.25 in); length 39.60 m (129 ft 10 in); height 10.16 m (33 ft 4 in); wing area 140.00 m² (1,507.00 sq ft)
Weights: empty equipped 36000 kg (79,365 lb); maximum take-off 63500 kg (139,991 lb)
Performance: maximum level speed 'clean' at 6400 m (21,000 ft) 389 kt (722 km/h; 448 mph); maximum cruising speed at 8230 m (27,000 ft) 330 kt (611 km/h; 380 mph); patrol speed at 600 m (1,985 ft) 215 kt (400 km/h; 248 mph); take-off run 1300 m (4,265 ft) at maximum take-off weight; ferry range 3,885 nm (7200 km; 4,474 miles); patrol endurance 12 hours
Armament: two internal weapons bays for homing torpedoes, nuclear and conventional depth charges, and sonobuoys

Ilyushin Il-76/-78/A-50 *Military freighter/tanker/AEW & C platform*

Red Star AWACS: the A-50 'Mainstay' is Russia's airborne early warning version of the Il-76 heavylift transport. Most are located at the Pechora base in the polar region.

The **Ilyushin Il-76** was developed as a successor to the An-12 for both Aeroflot and the Soviet air force. Larger, heavier and more powerful than the contemporary C-141, the Il-76 uses extensive high lift devices, thrust reversers and a high flotation undercarriage to achieve much better short- and rough-field performance. The cargo hold is fully pressurised and can be quickly reconfigured by using interchangeable passenger, freight or air ambulance modules. The first prototype **'Candid-A'** made its first flight on 25 March 1971 and a development squadron was in service by 1974. Series production began in 1975 and more than 750 had been built by the beginning of 1993.

'Candid-B'

'Candid-A' sub-types comprise the **Il-76T** featuring additional fuel tankage, **Il-76TD** with uprated Soloviev D-30KP-1 turbofans for improved 'hot-and-high' performance, four **Il-76LL** engine testbeds and the sole **Il-76DMP** firebomber conversion. **'Candid-B'** is used to identify dedicated military versions of the Il-76, which can be externally identified by a manned tail gun turret. The first dedicated military variant was the **Il-76M**, which was equivalent to the Il-76T. The designation **Il-76MD** is used for D-30KP-1-powered aircraft. Foreign operators are India, Iraq, Libya and Syria.

Specialised tanker and AWACS variants

The **Il-78M 'Midas'** is a three-point tanker fitted with three UPAZ external refuelling HDUs, one under each wing and one mounted on the port side of the rear fuselage. Internally the Il-78M has two pallet-mounted tanks in the hold; each contains 35 tonnes of fuel.

The **A-50 'Mainstay'** was developed as an AEW and AWACS platform. It has a rotodome above the fuselage, with the nose glazing and tail turret removed and replaced by further radomes. The A-50's performance is roughly comparable to Boeing's E-3 Sentry, with an inferior absolute detection range, but a (claimed) superior ability to discriminate against ground clutter. Related AEW variants are the **Il-976** and Iraq's indigenously developed **Adnan**.

Most Russian 'Midas' tankers wear Aeroflot colours. This early Il-78 has only one fuselage-mounted refuelling pod.

Ilyushin Il-76M 'Candid-A'

Specification: Ilyushin Il-76M 'Candid-B'
Powerplant: four PNPP 'Aviadvigatel' (Soloviev) D-30KP turbofans each rated at 117.68 kN (26,455 lb st)
Dimensions: wing span 50.50 m (165 ft 8 in); length 46.59 m (152 ft 10.25 in); height 14.76 m (48 ft 5 in); wing area 300.00 m² (3,229.28 sq ft)
Weights: maximum take-off 170000 kg (374,780 lb); maximum payload 40000 kg (88,183 lb)
Performance: maximum level speed 'clean' at optimum altitude 459 kt (850 km/h; 528 mph); cruising speed between 9000 and 12000 m (29,530 and 39,370 ft) between 405 and 432 kt (750 and 800 km/h; 466 and 497 mph); absolute ceiling about 15500 m (50,855 ft); take-off run 850 m (2,790 ft) at MTOW; landing run 450 m (1,475 ft) at normal landing weight; ferry range 3,617 nm (6700 km; 4,163 miles); range 2,698 nm (5000 km; 3,107 miles) with maximum payload
Armament: rear turret with gunner and twin radar-directed NR-23 23-mm cannon

Kamov Ka-50 'Hokum'

The extraordinary Ka-50 'Hokum' has been chosen for production for the Russian army in the anti-tank role.

The **Ka-50 'Hokum'** was developed as a rival to the Mil Mi-28 in the competition to provide a new battlefield helicopter for the Soviet armed forces. Realising that it would be difficult to achieve AH-64 levels of performance with existing Soviet technology and equipment, Kamov followed an individualistic course, retaining its trademark coaxial contra-rotating rotor configuration. This was felt to endow much better agility, reduced vulnerability and a more compact airframe. Kamov anticipated problems in keeping weight down, since heavier armour, more powerful armament and advanced sensors would all be required. Kamov also decided to design a single-seat helicopter, using their experience of sophisticated autohover systems on their naval helicopters.

Single-cockpit and ejection seat

The single-pilot cockpit was successfully demonstrated on the testbench, and in a modified Ka-29TB. A novel feature is the pilot's Severin/Zvezda K-37 ejection seat. The ejection sequence begins with automatic rotor separation, then jettisons the doors before a rocket pack drags the seat from the helicopter. The **V.80** prototype made its maiden flight on 27 July 1982. The competitive evaluation ended in October 1986 and the Ka-50 was reportedly selected in preference to the Mil Mi-28, although small batches of each are being procured for evaluation.

Missile and gun armament

The tube-launched, laser-beam-riding Vikhr (NATO AT-9 'Whirlwind') missile forms the Ka-50's primary armament. Sixteen can be carried, augmented by the built-in 30-mm cannon. Developed for the BMP AFV, the gun has variable rates of fire and selective feed from two 250-round ammunition boxes. The gun is installed on the starboard side of the fuselage below the wing root and is electro-hydraulically driven and can be traversed through 30° in elevation, and can also move 15° in azimuth. Combat survivability is enhanced by the IR suppressors in the exhaust assemblies, the heavily armoured pressurised cockpit, the foam-filled, self-sealing fuel tanks. Wingtip pods house chaff/flare dispensers.

The Werewolf retains Kamov's trademark coaxial contra-rotating rotor configuration. Single-seat operation is made possible by the use of sophisticated automated systems.

Kamov Ka-50 Werewolf

Specification: Kamov Ka-50 Werewolf 'Hokum'
Powerplant: two Klimov (Isotov) TV3-117VK turboshafts each rated at 1660 kW (2,226 shp)
Dimensions: rotor diameter, each 14.50 m (45 ft 6.9 in); length overall, with rotors turning 16.00 m (52 ft 5.9 in), and fuselage excluding probe and gun 13.50 m (44 ft 3.5 in); height 5.40 m (17 ft 8.6 in); rotor disc area, total 330.26 m² (3,555.00 sq ft)
Weights: maximum take-off 7500 kg (16,534 lb)
Performance: maximum level speed 'clean' at optimum altitude 188 kt (350 km/h; 217 mph) maximum vertical rate of climb at 2500 m (8,200 ft) 600 m (1,969 ft) per minute; hovering ceiling 4000 m (13,125 ft) out of ground effect; combat radius about 135 nm (250 km; 155 miles)
Armament: one 2A42 30-mm cannon in starboard fuselage side with two 250-round drums; primary armament of up to sixteen AT-9 Vikhr anti-armour missiles; other ordnance includes AS-12 'Kegler' guided missiles, 80 unguided S-8 80-mm rockets in four B-8 pods, AAMs, ARMs and 23-mm gun pods

Kawasaki T-4

An indigenous product of the Japanese aviation industry, the T-4 follows the basic configuration adopted for many other advanced trainer designs.

The **Kawasaki T-4** was developed as an intermediate jet trainer to replace the Lockheed T-33 and indigenous Fuji T-1A/B. Design studies were completed in 1982 and four prototypes (designated **XT-4**) were funded in 1984, the first of these making the type's maiden flight on 29 July 1985. The T-4 is of entirely conventional design, featuring high subsonic manoeuvrability and docile handling characteristics. It shares a similar high-wing configuration to the Dassault-Dornier Alpha Jet, with pronounced anhedral and large 'dog tooth' leading edges. Visibility for both instructor and pupil is excellent, with a frameless wrap-around windscreen and a one-piece canopy. For its secondary liaison role, the T-4 has a baggage compartment fitted in the centre fuselage with external access via a door in the port side.

Indigenous manufacture

The T-4 is a collaborative venture, in which Fuji builds the rear fuselage, supercritical section wings and tail unit, and Mitsubishi the centre fuselage and air intakes. Kawasaki builds only the forward fuselage, but is responsible for final assembly and flight test. Virtually all components are indigenously built, and most are locally designed, including the 16.37-kN (3,680-lb st) Ishikawajima-Harima F3-IHI-30 turbofan engines. A single underwing pylon on each side can accommodate a 450-litre (99-Imp gal) drop tank, and a centreline pylon can be fitted for a target towing winch, air sampling pod, ECM pod or chaff dispenser.

Service

Production deliveries began in September 1988, to meet a requirement for some 200 aircraft. By mid-1992 126 had been ordered, and the T-4 was in service with Nos 31 and 32 Squadrons of No. 1 Air Wing at Hamamatsu, with some operational squadrons and wings as a hack, sometimes wearing a camouflage colour scheme. An advanced version, presumably with more powerful engines and advanced avionics, has been offered to the JASDF to replace the ageing and uneconomical Mitsubishi T-2.

In addition to undertaking an advanced training role with the JASDF, the T-4 is used for liaison, target-towing and utility duties. An aft baggage compartment is fitted for liaison tasks.

Kawasaki T-4

Specification: Kawasaki T-4
Powerplant: two Ishikawajima-Harima F3-IHI-30 turbofans each rated at 16.32 kN (3,671 lb st)
Dimensions: wing span 9.94 m (32 ft 7.5 in); length 13.00 m (42 ft 8 in); height 4.60 m (15 ft 1.25 in); wing area 21.00 m² (226.05 sq ft)
Weights: empty 3700 kg (8,157 lb); normal take-off 5500 kg (12,125 lb); maximum take-off 7500 kg (16,534 lb)
Performance: maximum level speed 'clean' at 10975 m (36,000 ft) 516 kt (956 km/h; 594 mph) and at sea level 560 kt (1038 km/h; 645 mph); cruising speed at 10975 m (36,000 ft) 430 kt (797 km/h; 495 mph); maximum rate of climb at sea level 3048 m (10,000 ft) per minute; service ceiling 15240 m (50,000 ft); take-off run 549 m (1,800 ft) at normal take-off weight; landing run 671 m (2,200 ft) at normal landing weight; ferry range 900 nm (1668 km; 1,036 miles) with drop tanks; standard range 700 nm (1297 km; 806 miles) with standard fuel
Armament: no built-in armament; structural provision for up to 2000 kg (4,409 lb) of ordnance

Lockheed C-5 Galaxy

For many years the world's largest aircraft, the C-5 remains the USAF's most valuable transport asset. Strategic airlift is its raison d'être and over 120 are left in service.

The **Lockheed C-5 Galaxy** heavy logistics transport originated from a 1963 USAF CX-HLS (Cargo Experimental-Heavy Logistics System) requirement for a capability to carry a 113400-kg (250,000-lb) payload over 4828 km (3,000 miles) without air refuelling. The resulting design incorporates a high-wing, T-tailed configuration, powered by four underwing podded TF39 turbofans. Key to the C-5's mission is its cavernous interior and 'roll on/roll off' capability, with access to the vast cargo bay at both front and rear via an upward-lifting visor nose and standard rear clamshell doors. The **C-5A** first flew on 30 June 1968 and operational C-5s were delivered between 17 December 1969 and May 1973. The C-5A suffered initially from wing crack problems and cost overruns, but has served well after a teething period. Seventy-seven C-5As underwent a re-winging programme from 1981 to 1987.

Improved model

In the mid-1980s, the production line was reopened to meet an urgent USAF demand for additional heavy airlift capacity. Fifty **C-5B** models were built, essentially similar to the C-5A, but incorporating modifications and improvements resulting from operations with the C-5A. The B model dispensed with the C-5A's complex crosswind main landing gear and introduced an improved AFCS (automated flight control system). The first production C-5B was delivered on 8 January 1986 and deliveries were completed by 1989.

Typical mission payloads

Typical loads include two M1A1 Abrams MBTs, four M551 Sheridan light tanks and one HMMVW, 16 ¾-ton trucks, 10 LAV-25s, or a CH-47. Although not usually assigned airdrop duties, the Galaxy can also drop paratroopers. C-5A/Bs serve with six active, four Reserve, and one ANG squadron. Galxies have served in airlifts supporting US operations in Vietnam, Israel (October 1973 War), and Desert Shield/Storm (1990-91), during which they flew 42 per cent of cargo and 18.6 per cent of passenger missions. The C-5 has no immediate replacement and will serve until at least 2010.

The C-5's prodigious capacity and global reach (with inflight refuelling) make it vital to the US rapid deployment forces.

Lockheed C-5B Galaxy

Specification: Lockheed C-5B Galaxy
Powerplant: four General Electric TF39-GE-1C turbofans each rated at 191.27 kN (43,000 lb st)
Dimensions: wing span 67.88 m (222 ft 8.5 in); length 75.54 m (247 ft 10 in); height 19.85 m (65 ft 1.5 in); wing area 575.98 m² (6,200.00 sq ft)
Weights: operating empty 169643 kg (374,000 lb); maximum take-off 379657 kg (837,000 lb); maximum payload 118387 kg (261,000 lb)
Performance: maximum level speed at 7620 m (25,000 ft) 496 kt (919 km/h; 571 mph); maximum cruising speed at 7620 m (25,000 ft) between 460 and 480 kt (888 and 908 km/h; 552 and 564 mph); maximum rate of climb at sea level 525 m (1,725 ft) per minute; service ceiling 10895 m (35,750 ft) at 278960 kg (615,000 lb); take-off run 2530 m (8,300 ft) at MTOW; take-off distance to 15 m (50 ft) 2987 m (9,800 ft) at MTOW; landing distance from 15 m (50 ft) 1164 m (3,820 ft) at maximum landing weight; landing run 725 m (2,380 ft) at maximum landing weight; range 5,618 nm (10411 km; 6,469 miles) with maximum fuel or 2,982 nm (5526 km; 3,434 miles) with maximum payload

Lockheed C-130 Hercules

The world's premier transport, Lockheed's C-130 has achieved an unequalled longevity and popularity, and serves with over 60 other nations. Large numbers remain in US service.

The **Lockheed C-130 Hercules** is the West's most widely used and versatile military transport. Over 2,000 have been produced, serving with 64 countries. It features an unobstructed and fully pressurised cargo compartment and can carry a 19686-kg (43,400-lb) payload over 2298 km (1,428 miles). In Somalia, C-130s carrying 9980 kg (22,000 lb) have routinely landed in 930 m (3,000 ft). Typical loads include five HMMWVs, five 4000-kg (8,818-lb) freight containers, or three Land Rovers and two trailers. The C-130E carries 64 paratroopers to a radius of up to 1142 km (710 miles).

Early variants

The **YC-130** first flew on 23 August 1954, and was followed by the production **C-130A** in April 1955. USAF deliveries began that December. The **C-130B** featured T56-A-1A engines, increased fuel capacity and four-bladed propellers. In 1961, production changed to the **C-130E** variant, introducing T56-A-7 engines with increased power to improve 'hot-and-high' performance, and larger external underwing tanks. It also had strengthened spars, thicker skins and a reinforced undercarriage. Sub-types include the **TC-130** trainer, the **JC-130** and **NC-130** test aircraft, the recce **RC-130**, the **VC-130** VIP transport, and the **WC-130** weather ship.

C-130H

The current basic **C-130H** was developed for export. It first flew in November 1964 and the first production transports were delivered to the RNZAF in 1965. The USAF received its first H in April 1975. The C-130H features a strengthened wing box, improved brakes, new avionics and T56-A-15 engines. Variants include the **C-130H(AEH)** airborne hospital and the **C-130H-MP** maritime patroller. The RAF's **C-130K** (**Hercules C.Mk 1**) has British avionics and equipment. Thirty were 'stretched' by 4.5 m (15 ft) to become **C.Mk 3**s, increasing capacity from 92 to 128 troops.

Lockheed is developing the much-improved, high-technology **C-130J**. Powered by Allison 2100 turboprops driving six-bladed, composite propellers, this will have a two-man cockpit with advanced flat-panel liquid crystal displays.

The RAF is a major C-130 user. This C.Mk 1 celebrated 25 years of operations in 1992 and serves alongside stretched C.Mk 3s.

Lockheed Hercules C.Mk 3P (C.Mk 1P side view)

Specification: Lockheed C-130F Hercules
Powerplant: four Allison T56-A-7 turboprops each rated at 3020 ekW (4,050 ehp)
Dimensions: wing span 40.41 m (132 ft 7 in); length 29.79 m (97 ft 9 in); height 11.66 m (38 ft 3 in); wing area 161.12 m² (1,745.00 sq ft)
Weights: empty equipped 31434 kg (69,300 lb); maximum take-off 61236 kg (135,000 lb); maximum payload 16194 kg (35,700 lb)
Performance: maximum cruising speed at 9145 m (30,000 ft) 321 kt (595 km/h; 370 mph); maximum rate of climb at sea level 610 m (2,000 ft) per minute; service ceiling 10365 m (34,000 ft); take-off distance to 15 m (50 ft) 1311 m (4,300 ft) at MTOEW range 4,210 nm (7802 km; 4,848 miles) with maximum fuel or 1,910 nm (3539 km; 2,199 miles) with maximum payload

Lockheed C-130 special variants

The Hercules has bettered the 12 labours of its mythical namesake with a myriad of different roles. The latest 'special' model is the AC-130U gunship, armed with a new 25-mm gun.

The versatility of the basic C-130 transport has led to conversions for specialised missions. The initial **AC-130A** gunship version introduced specialised sensors, including a searchlight, ignition sensors, a FLIR and an LLLTV camera. Successive **AC-130E** and **AC-130H** variants introduced improved armour, avionics, engines and armament (a 105-mm howitzer and 40-mm Bofors cannon). The current **AC-130U** version has a single 25-mm cannon replacing the two 20-mm M61s of previous Spectres. Sensors include AN/APG-180 radar, all-light-level TV, and ALQ-172 jammer package.

Several variants carry the designation **EC-130E**, undertaking Sigint, Comint and Elint missions. Three are operational today. The first is the Airborne Battlefield Command and Control Center (**ABCCC**) variant. The **EC-130E(CL)** flies 'Senior Scout' Elint missions, the **EC-130E(RR)** 'Rivet Rider' undertakes psy-war tasks and the **EC-130H** 'Compass Call' fulfils stand-off jamming duties.

Long-range SAR variants include the **HC-130H**, with a large radar above the forward fuselage, the similar **HC-130H-7** for the USCG, and the **HC-130N** and **HC-130P** (HDU-equipped and used to refuel rescue helicopters). The **HC-130H(N)** has modernised avionics and lacks Fulton gear, but is equipped with underwing HDUs. Dedicated tankers include the **KC-130B**, **KC-130H**, and **KC-130T** and the stretched **KC-130T-30**. Not in US service, the **KC-130H** (similar to the **KC-130R**) has been widely exported.

Special forces support

Initial **MC-130E Combat Talon** Special Forces variants served in Vietnam, equipped with Fulton STAR recovery gear for retrieval of downed aircrew/recce troops. They featured IFR capability, uprated engines, and INSs. Current sub-variants comprise the Fulton-equipped **MC-130E-C**, the **MC-130-Y** with a 'Pinocchio' nose, and the Sigint-tasked **MC-130E-S**. All MC-130Es have a radar for weather avoidance, long-range navigation (with terrain-following) and a FLIR for low-level night/adverse weather penetration. The latest **MC-130H Combat Talon II** features a new APQ-170 radar and an improved FLIR underneath.

Blade antennas on the fin leading edge and under the wings identify this 'special' Hercules as the EC-130E 'Rivet Rider'.

Lockheed AC-130H Spectre

Specification: Lockheed AC-130U Hercules
Powerplant: four Allison T56-A-15 turboprops each rated at 3362 ekW (4,508 ehp)
Dimensions: wing span 40.41 m (132 ft 7 in); length 29.79 m (97 ft 9 in); height 11.66 m (38 ft 3 in); wing area 162.12 m² (1,745.00 sq ft)
Weights: operating empty 34356 kg (75,743 lb); normal take-off 70310 kg (155,000 lb); maximum take-off 79380 kg (175,000 lb)
Performance: maximum cruising speed at optimum altitude 325 kt (602 km; 374 mph); economical cruising speed at optimum altitude 300 kt (556 km/h; 345 mph); maximum rate of climb at sea level 579 m (1,900 ft) per minute; service ceiling at 58967 kg (130,000 lb) 10060 m (33,000 ft); take-off distance to 15 m (50 ft) 1573 m (5,160 ft) at MTOW; landing distance from 15 m (50 ft) 838 m (2,750 ft) at 58967 kg (130,000 lb); range 4,250 nm (7876 km; 4,894 miles) with maximum fuel and a 7081-kg (15,611-lb) payload, or 2,046 nm (3791 km; 2,356 miles) with maximum payload

Lockheed C-141 StarLifter

Air Mobility Command's C-141B fleet has been repainted in 'proud grey'. Stretching the C-141 gained the USAF valuable extra lift capacity and greatly enhanced productivity.

First flown on 17 December 1963, the **Lockheed C-141A StarLifter** has provided the USAF with a fast and versatile long-range jet transport since it entered service in 1964. The design features a fuselage of similar cross-section to the C-130, two large clamshell doors and a rear ramp which could be opened in flight for air-dropping. Swept wings were adopted for high-speed cruise, with powerful high-lift devices provided for good low-speed field performance. Power came from four podded TF33 fuel-efficient turbofans, and all fuel was housed integral wing tanks. The aircraft commenced squadron operations with MAC in April 1965, supplying the war effort in Vietnam.

Stretched, improved C-141B

It soon became apparent that the C-141A's maximum payload of 32136 kg (70,847 lb) (or 41731 kg/92,000 lb on C-141As configured to carry LGM-30 Minuteman ICBMs) was rarely achieved, the aircraft 'bulking out' before its weight limit was approached. During the 1970s, the entire fleet (minus four **NC-141A** test aircraft) was cycled through a programme to bring all 270 aircraft to **C-141B** standard. This added a 7.11-m (23-ft 4-in) fuselage stretch and IFR receptacle above the cockpit for true global airlift capacity. Overall cargo capacity has been increased by over 30 per cent, providing the equivalent of 90 new C-141s in terms of capacity at low relative cost. The prototype **YC-141B** made its first flight on 24 March 1977 and Lockheed completed the final C-141B in June 1982.

Typical payloads

Palletised seats can be fitted for 166 passengers, while by using canvas seats some 205 passengers or 168 paratroops can be carried. For medevac missions, the C-141B can carry 103 litter patients and 113 walking wounded. It can also carry a Sheridan tank, an AH-1 helicopter or five HMMWVs. Thirteen standard cargo pallets can be admitted, and other loads can include aircraft engines, food supplies, fuel drums or nuclear weapons. Of current major concern is the rapid ageing of the C-141 fleet.

This C-141B displays the previous 'lizard' camouflage scheme. Six AMC wings fly the StarLifter on general airlift duties.

Lockheed C-141B StarLifter

Specification: Lockheed C-141B StarLifter
Powerplant: four Pratt & Whitney TF33-P-7 turbofans each rated at 93.41 kN (21,000 lb st)
Dimensions: wing span 48.74 m (159 ft 11 in); length 51.29 m (168 ft 3.5 in); height 11.96 m (39 ft 3 in); wing area 299.83 m² (3,228.00 sq ft)
Weights: operating empty 67186 kg (148,120 lb); maximum take-off 155580 kg (343,000 lb); maximum payload 41222 kg (90,880 lb) at 2.25 g or 32026 kg (70,605 lb) at 2.5 g
Performance: maximum cruising speed at high altitude 492 kt (910 km/h; 566 mph); economical cruising speed at high altitude 430 kt (796 km/h; 495 mph); maximum rate of climb at sea level 890 m (2,920 ft) per minute; service ceiling 12680 m (41,600 ft); take-off distance to 15 m (50 ft) 1768 m (5,800 ft) at maximum take-off weight; landing distance from 15 m (50 ft) 1128 m (3,700 ft) at normal landing weight; ferry range 5,550 nm (10280 km; 6,390 miles); range 2,550 nm (4725 km; 2,936 miles) with maximum payload

Lockheed F-16A/B Fighting Falcon

Pakistan is one of many F-16 export customers, receiving a total of 51 F-16A/Bs between 1983 and 1988. Its aircraft have scored kills against intruders along the Afghan border.

The **F-16 Fighting Falcon** is the most numerous fighter in the West and was originally conceived as a small light-weight 'no frills' fighter optimised for air-to-air combat. However, it has since evolved into a versatile and effective radar-equipped multi-role fighter with additional air-to-ground and all-weather navigation capability.

USAF service
First flown on 20 January 1974, the **YF-16** prototype was followed by FSD **F-16A** airframes in 1975 and the first FSD **F-16B** two-seater in 1977. Block 15 F-16A/Bs (also known as **MSIP I**) introduced the now-standard extended horizontal stabilator, or 'big tail'. The Multi-national Staged Improvement Program, adopted by Belgium, Denmark, the Netherlands, Norway and the US, upgrades earlier aircraft to Block 15 standard and also introduces improved APG-66 (V2A) radar and HUD. From October 1986 the USAF began conversion of 270 Block 15 F-16A/Bs to **ADF (Air Defense Fighter)** interceptor standard. This introduces a radar upgrade to improve small target detection and provide continuous-wave illumination, thereby giving AIM-7 Sparrow BVR missile capability. The ADF F-16 can carry up to six AIM-120, AIM-7 or AIM-9 AAMs.

Research programmes
F-16 research programmes include the **CCV** (Control-Configured Vehicle) with twin intake-mounted canards and **AFTI** (Advanced Fighter Technology Integration) feaaturing a triplex digital flight-control system, larger vertical canards and a thick dorsal spine. The SCAMP (Supersonic Cruise and Maneuvering Prototype), or **F-16XL**, was conceived to increase weapons capacity, range and penetration speed of the F-16. The F-16XL's fuselage was lengthened to 16.51 m (54 ft 2 in) and grafted to a 'cranked-arrow' delta wing. Following the first flight in July 1982, tests successfully demonstrated the validity of the design. The USAF abandoned development in the late 1980s and two aircraft were handed over to NASA, which operates them on supersonic laminar-flow flight tests.

Two-seat F-16Bs complement the single-seat F-16A force. They retain full combat capability with reduced fuel capacity.

Lockheed F-16ADF Fighting Falcon

Specification: General Dynamics F-16A Fighting Falcon
Powerplant: one Pratt & Whitney F100-P-100 turbofan rated at 65.26 kN (14,670 lb st) dry and 106.0 kN (23,830 lb st) with afterburning
Dimensions: wing span 10.00 m (32 ft 9.75 in) with tip-mounted AAMs; length 15.03 m (49 ft 4 in); height 5.01 m (16 ft 5.2 in); wing area 28.87 m^2 (300.00 sq ft)
Weights: operational empty 6607 kg (14,567 lb); typical combat take-off 10335 kg (22,785 lb); maximum take-off 14968 kg (33,000 lb)
Performance: maximum level speed 'clean' at 12190 m (40,000 ft) more than 1,146 kt (2124 km/h; 1,320 mph) and at sea level 795 kt (1472 km/h; 915 mph); maximum rate of climb at sea level more than 15240 m (50,000 ft) per minute; service ceiling more than 15240 m (50,000 ft); combat radius 295 nm (547 km; 340 miles) on a hi-lo-hi mission with six 454-kg (1,000-lb) bombs
Armament: one M61A1 Vulcan 20-mm cannon with 515 rounds; maximum ordnance 15,200 lb (6894 kg), including two AIM-9 Sidewinder missiles on wingtip launch rails

Lockheed F-16C/D Fighting Falcon

Identified by its longer dorsal spine and small blade antenna, the F-16C introduced improved avionics and systems. South Korea will eventually receive over 150 F-16C/Ds.

The **Lockheed F-16C/D** is basically an improved development of the F-16A/B series incorporating various structural, avionics and systems modifications. F-16C/D models are distinguished by an enlarged vertical fin base. Cockpit changes include a wide-angle HUD and an improved data display for HOTAS flying. Hughes' APG-68 multi-mode radar is installed with increased range, sharper resolution and expanded operating modes, and a weapons interface for AGM-65D and AMRAAM missiles. The first F-16C flew on 19 June 1984 and Block 25 aircraft entered production in July 1984. Blocks 32 and 42 aircraft introduced 106.05-kN (23,840-lb) thrust P&W F100-PW-200 engines. In addition, Block 30/32 aircraft are able to carry AGM-45 Shrike I and AGM-88A HARM ARMs, and AIM-120 AMRAAM. The US Navy's **F-16N** serves with adversary units and is essentially a small-inlet F-16C/D Block 30 aircraft with APG-66 radar, but without M61 cannon or missiles.

Night Falcon and Block 50 F-16s

From December 1988, Block 40/42 **Night Falcon**s introduced LANTIRN navigation and targeting pods, GPS navigation receiver, AGM-88B HARM II, APG-68V radar, digital flight controls, automatic terrain following, and (as a consequence) increased take-off weight, necessitating strengthened undercarriage. In December 1991, Block 50 and 52 aircraft were introduced with APG-68(V5) radar with improved memory and additional modes, new NVG-compatible GEC HUD, improved avionics computer and AN/ALE-47 chaff/flare dispenser. These latest F-16s are intended for the IPE (Improved Performance Engines) versions of GE and P&W engines, the F110-GE-229 and F100-PW-220, respectively.

Export F-16C/Ds

F-16C/Ds have been exported to, or ordered by, Bahrain, Egypt, Israel, Korea and Turkey. Licence manufacture of some F-16C/Ds is undertaken in Korea and in Turkey. Many F-16Ds delivered to Israel have been subsequently fitted with a bulged spine, housing unidentified indigenous avionics.

The F-16C/D is the USAF's standard fighter type and is used primarily for air-to-ground work.

Lockheed F-16C (F-16D side view)

Specification: Lockheed/General Dynamics F-16C Fighting Falcon
generally similar to the F-16A/B except in the following particulars
Powerplant: one GE F110-GE-100 turbofan rated at 122.77 kN (27,600 lb st) with afterburning
Dimensions: tailplane span 5.58 m (18 ft 3.75 in); height 5.09 m (16 ft 8.5 in)
Weights: empty 8663 kg (19,100 lb); typical combat take-off 9791 kg (21,585 lb); maximum take-off 11372 kg (25,071 lb) for an air-to-air mission without drop tanks or 19187 kg (42,300 lb) with maximum external load
Armament: one internal M61 Vulcan 20-mm cannon; maximum ordnance 6894 kg (15,200 lb), including Mk 20 Rockeye and CBU-87 cluster bombs, Mk 83 and Mk 84 227-kg (500-lb) and 454-kg (1,000-lb) bombs, AGM-65 Maverick missiles, and GBU-10 and GBU-15 guided weapons

Lockheed F-104 Starfighter

Italy's ASA upgrade for 147 F-104S interceptors adds a 20-mm Vulcan cannon, and compatibility with the indigenous medium-range Aspide (modified AIM-7 Sparrow) 1A AAM.

Some 2,221 single-seat Starfighters were eventually built in the US, Canada, Europe and Japan in Lockheed's **F-104** programme, which started in the early 1950s at the renowned 'Skunk Works'. The first of two XJ65-powered **XF-104** prototypes started flying on 18 February 1954, followed by 17 pre-production **YF-104A**s with definitive GE J79 turbojets.

European licence-production

Extensive modifications and installation of multi-role nav/attack systems and the improved 70.28-kN (15,800-lb) J79-GE-11A secured for Lockheed in late 1958 an initial West German contract for the new **F-104G** Starfighter. This involved a massive European licence-production programme by Belgium, Germany, Italy and the Netherlands, with additional industrial participation from Canada, which also selected the **CF-104** as its next-generation fighter. Production was also undertaken in Japan, where Mitsubishi built 210 Lockheed **F-104J**s for the JASDF. These were retired from front-line service in March 1986. Large numbers of surplus F-104s have been transferred to US allies and other NATO countries. Major recipients included Greece (at least 230), Norway (18), Taiwan (at least 75), and Turkey (266). In early 1994, small numbers of F-104G/Js were still operating in Taiwan and Turkey, although due for imminent replacement.

Italian F-104S interceptors

Final licence-production was undertaken in Italy (by Aeritalia) where a total of 246 **F-104S** interceptors was developed and built for the Italian air force (AMI), including 40 for Turkey. The F-104S was optimised for interception, rather than ground attack, duties. The ASA upgrade programme was launched in 1981 for 147 AMI F-104Ss. These introduced radar with look-down/shoot-down capability and compatibility with SARH medium-range Aspide AAMs, improved avionics, and AIM-9L AAMs. Due to delays in the introduction of the European Fighter Aircraft, further upgrades have been mooted for the F-104ASA.

Turkey is the largest F-104 operator, with over 200 in service. This fleet includes about 20 surviving F-104Ss of 40 supplied.

Lockheed F-104G Starfighter

Specification: Alenia (Aeritalia) F-104S ASA
Powerplant: one General Electric J79-GE-19 turbojet rated at 52.80 kN (11,870 lb st) dry and 79.62 kN (17,900 lb st) with afterburning
Dimensions: wing span 6.68 m (21 ft 11 in) without tip tanks; length 16.69 m (54 ft 9 in); height 4.11 m (13 ft 6 in); wing area 18.22 m² (196.10 sq ft)
Weights: empty 6760 kg (14,903 lb); normal take-off 9840 kg (21,693 lb); maximum take-off 14060 kg (30,996 lb)
Performance: maximum level speed 'clean' at 10975 m (36,000 ft) 1,259 kt (2333 km/h; 1,450 mph), and at sea level 790 kt (1464 km/h; 910 mph); maximum rate of climb at sea level 16764 m (55,000 ft) per minute; climb to 10670 m (35,000 ft) in 1 minute 20 seconds; service ceiling 17680 m (58,000 ft); combat radius 673 nm (1247 km; 775 miles) with maximum fuel
Armament: one M61 Vulcan 20-mm cannon, maximum ordnance 3400 kg (7,495 lb), typical intercept load of two Aspide medium-range AAMs and two AIM-9L short-range AAMs

Lockheed F-117 Night Hawk

USA
Stealthy attack aircraft

The F-117 is used in precision attacks against high-value ground targets. In the Gulf War, they spearheaded the attack effort against air defence and communications centres.

The effects of the Vietnam and Yom Kippur Wars spurred a DARPA request in 1974 for development of a stealthy aircraft, using a mix of RAM, radar-reflective internal structure and a similarly 'reflective' configuration to dramatically decrease radar cross-section (RCS). Lockheed's 'Skunk Works' developed two sub-scale 'Have Blue' technology demonstrators, which first flew in 1977. Both utilised a faceted structure to reduce RCS and, although both aircraft were lost, experience gained was sufficient to win Lockheed a contract to develop a full-scale operational tactical fighter, awarded on 16 November 1978 under the 'Senior Trend' codename. Introducing a revised, outboard-canted tailfin configuration, the first of five **F-117** FSD prototypes flew on 18 June 1981.

Operations

As production of 59 aircraft continued at a low rate, the USAF began establishing a base at Tonopah Test Range in Nevada. In October 1983, the first unit was declared operational, undertaking only night flights until November 1988. There is no two-seat trainer version; instead, pilots rely on very sophisticated simulators and T-38s for proficiency training. The F-117's baptism of fire in Panama was overshadowed by the type's contribution to Desert Storm, when an eventual total of 42 aircraft flew from Saudi Arabia on nightly missions against Iraq and occupied Kuwait.

The F-117 is used for attacks against 'highly leveraged' targets such as communications and command centres, air defence sector centres, key bridges and airfields. After the target has been acquired by the FLIR and locked into the weapons computer, it is passed over to the DLIR sensor mounted under the nose. Laser-guided bombs are released at a computed point, and a laser (boresighted with the DLIR) is used to designate the impact point.

A post-Gulf War Offensive Capability Improvement Program (OCIP) aims to reduce cockpit workload, adding two colour MFDs, a moving map display and auto-throttles. Further improvements will add a new IR acquisition and designation sensor (replacing FLIR and DLIR).

After years of operating in secrecy, the USAF's F-117 force is now based at Holloman AFB, New Mexico.

Lockheed F-117A Night Hawk

Specification: Lockheed F-117A Night Hawk
Powerplant: two General Electric F404-GE-F1D2 non-afterburning turbofans each rated at 48.04 kN (10,800 lb st)
Dimensions: wing span 13.20 m (43 ft 4 in); length 20.08 m (65 ft 11 in); height 3.78 m (12 ft 5 in); approximate wing area 105.9 m² (1,140.00 sq ft)
Weights: empty about 13608 kg (30,000 lb); maximum take-off 23814 kg (52,500 lb)
Performance: maximum level speed 'clean' at high altitude possibly more than Mach 1; normal maximum operating speed at optimum altitude Mach 0.9; combat radius about 600 nm (1112 km; 691 miles) with maximum ordnance
Armament: maximum ordnance 2268 kg (5,000 lb); standard ordnance comprises GBU-10 Paveway II or GBU-27 Paveway III 2,000-lb LGBs, each with either standard Mk 84 or the BLU-109 penetration warheads; weapons bay may be configured for AAMs, gun pods, AGM-65 Maverick ASMs, and AGM-88 HARMs

70

Lockheed P-3 Orion

US Navy ASW Orions wear this very muted all-over grey scheme. The P-3C is the primary shore-based anti-submarine platform; this Orion serves with VP-16, based at Jacksonville.

Based on the Lockheed L-188 Electra medium-range passenger airliner, the Lockheed P-3 Orion was developed to meet a 1957 US Navy requirement for a new anti-submarine aircraft to replace the Lockheed P-2 Neptune. An initial batch of seven **P-3A**s was ordered and the Orion entered service in mid-1962. The following **P-3B** variant introduced uprated Allison T56-A-14 engines, higher weights and provision for AGM-12 Bullpup ASMs. The **Lockheed P-3C Orion** variant entered service in 1969 and remains the US Navy's primary land-based, ASW patrol aircraft. Retaining the P-3B's airframe and powerplant, it uses a comprehensive package of ASW detection equipment, including a search radar, a MAD, a directional acoustics-frequency analysis and recording system, and a sonar tape recorder.

Lockheed P-3C Orion

Specialised variants

The US Navy employs a fleet of specially-modified Orions to perform the Elint-gathering role as the **EP-3E 'Aries'**. The **EP-3J** is an EW jamming trainer fitted with internally- and pod-mounted jamming equipment. Five Orions are used for range support work, comprising two **EP-3A SMILS** (Sono-buoy Missile Impact Locating System) aircraft and three **RP-3A (EATS)** (Extended Area Test System) aircraft used for accurate tracking and instrumentation of missile tests. Further variants include the oceanographic reconnaissance **RP-3A**, weather reconnaissance **WP-3A/D**, **VP-3A** executive transport, **TP-3A** aircrew trainer, **UP-3A** utility transport and **NP-3A/B** trials aircraft. The **P-3 AEW&C** featured an airborne early warning radar in a dorsal rotodome, and was used by the US Customs Service on anti-drug patrols.

Canadian Aurora

The **CP-140 Aurora** is the version of the Orion purchased in 1976 by the Canadian Armed Forces. It is configured internally to Canadian requirements, and is equipped with an avionics system based on that of the S-3A Viking. The **CP-140A Arcturus** is a special-duty variant of the CP-140 used for environmental, Arctic and fishery patrols.

The JMSDF has ordered a total of 99 Kawasaki-built P-3Cs (to Update II and III standards) to equip seven ASW squadrons.

Specification: Lockheed P-3C Orion
Powerplant: four Allison T56-A-14 turboprops each rated at 3661 ekW (4,910 ehp)
Dimensions: wing span 30.37 m (99 ft 8 in); length 35.61 m (116 ft 10 in); height 10.27 m (33 ft 8.5 in); wing area 120.77 m^2 (1,300.00 sq ft)
Weights: empty 27890 kg (61,491 lb); normal take-off 61235 kg (135,000 lb); maximum take-off 64410 kg (142,000 lb)
Performance: maximum level speed 'clean' at 4575 m (15,000 ft) 411 kt (761 km/h; 473 mph); economical cruising speed at 7620 m (25,000 ft) 328 kt (608 km/h; 378 mph); patrol speed at 457 m (1,500 ft) 206 kt (381 km/h; 237 mph); maximum rate of climb at 457 m (1,500 ft) 594 m (1,950 ft) per minute; service ceiling 8625 m (28,300 ft)
Armament: maximum expendable load 9072 kg (20,000 lb); 10 stores stations and internal weapons bay for mines, torpedoes, destructors, nuclear depth charges, conventional bombs, practice bombs and rockets; search stores include AN/ALQ-78 ESM pods, sonobuoys, smoke markers and parachute flares

Lockheed S-3 Viking

Each carrier air wing deploys a single ASW Viking squadron to fly the outer zone defence mission. The S-3B model features improved radar and acoustic processing capability.

Lockheed S-3A Viking (ES-3A side view)

Developed to counter Soviet quiet, deep-diving nuclear submarines, the **Lockheed S-3 Viking** is the US Navy's standard carrier-based, fixed-wing ASW aircraft. It is a conventional high-wing, twin-turbofan aircraft with accommodation for four crew (comprising pilot, co-pilot, tactical co-ordinator and acoustic sensor operator). The first of eight service-test **YS-3A**s made its maiden flight on 21 January 1972, and was followed by 179 production **S-3A**s equipped with AN/APS-116 search radar and OR-89 FLIR. The heart of the ASW suite is an AN/ASQ-81 MAD sensor housed in a retractable tailboom. The Viking carries 60 sonobuoys in its aft fuselage and has a ventral bomb bay and wing hardpoints for bombs, torpedoes or depth charges. The S-3A entered fleet service in July 1974.

The current **S-3B** variant is the result of a 1981 upgrade programme, which adds improved acoustic processing, expanded ESM coverage, increased radar processing capabilities, a new sonobuoy receiver system, and provision for AGM-84 Harpoon ASMs. It can be distinguished by a small chaff dispenser located on its aft fuselage. Nearly all surviving S-3As have been upgraded to S-3B standard.

Specialised Viking variants

Lockheed has produced four **US-3A** COD aircraft. These complement Grumman C-2As and are stripped of ASW equipment. The **KS-3A** dedicated tanker variant has not been produced, although operational S-3s have been adapted as part-time tankers with the same refuelling store.

First flying in May 1991, the **ES-3A** is a carrier-based Elint aircraft modified from the S-3A with OTH surveillance equipment. The co-pilot position is replaced by a third sensor station and the bomb bays have been modified to accommodate avionics. The ES-3A has a new radome, direction-finding antenna, and other equipment in a dorsal 'shoulder' on the fuselage, an array of seven receiving antennas on its underfuselage, cone-shaped omnidirectional Elint antenna on each side of the rear fuselage, and wingtip AN/ALR-76 ESM antennas. Two USN squadrons are to deploy eight ES-3As each on carriers, usually in pairs.

Fully equipped to hunt subs, the S-3B also has a kill capability in the form of destructors, mines and AGM-84 Harpoon missiles.

Specification: Lockheed S-3A Viking
Powerplant: two General Electric TF34-GE-2 turbofans each rated at 41.26 kN (9,275 lb st)
Dimensions: wing span 20.93 m (68 ft 8 in); length overall 16.26 m (53 ft 4 in) height overall 6.93 m (22 ft 9 in); wing area 55.56 m² (598.00 sq ft)
Weights: empty 12088 kg (26,650 lb); normal take-off 19277 kg (42,500 lb); maximum take-off 23832 kg (52,540 lb)
Performance: maximum level speed 'clean' at sea level 439 kt (814 km/h; 506 mph); patrol speed at optimum altitude 160 kt (296 km/h; 184 mph); maximum rate of climb at sea level over 1280 m (4,200 ft) per minute; service ceiling over 10670 m (35,000 ft); operational radius over 945 nm (1751 km; 1,088 miles); endurance 7 hours 30 minutes
Armament: (weapons bay) four Mk 46 torpedoes, or four Mk 36 destructors, or four Mk 82 bombs, or two Mk 57 or four Mk 54 depth bombs, or four Mk 53 mines: (underwing pylons), flare launchers, mines, cluster bombs and rockets

Lockheed U-2R

The U-2R reconaissance platform undertakes high-altitude all-weather, day-or-night battlefield surveillance.

Lockheed U-2R

Early variants of the **Lockheed U-2** had successful careers as high-altitude reconnaissance platforms. However, these models were airframe-limited, and Lockheed developed the improved **U-2R**, which first flew in August 1967. It was larger overall, offering greatly improved range/ payload, endurance and handling characteristics. Twelve early U-2Rs served with the USAF (in Vietnam) and with the CIA. In 1979 the production line reopened to provide 37 new airframes for the **TR-1A** programme, using the U-2R as a platform for the ASARS-2 battlefield surveillance radar, PLSS radar location system; and for Sigint-gathering equipment. The batch also included two **TR-1B**s and one **U-2RT**. All three two-seat trainers are identical, and the TR-1s have adopted the U-2R designations. Finally, two U-2Rs were built as **ER-2** NASA earth resources monitoring aircraft.

Sensor options

The U-2R resembles a powered glider in configuration. The retractable main bicycle undercarriage is supplemented by plug-in detachable 'pogo' outriggers. The wingtips incorporate skids, above which are RWRs. Sensors are carried in the nose, a large 'Q-bay' for cameras (behind the cockpit), smaller bays along the lower fuselage and in two removable wing 'super pods'. Sensors include Comint and Elint recorders, imaging radars, radar locators and high resolution cameras. A common fit is the Tactical Reconnaissance System (TRS), comprising ASARS-2 nose radar, SLAR, and a large 'farm' of Sigint antennas on the 'super pods' and rear fuselage. Recorded intelligence can be transmitted via datalink to ground stations, and at least three aircraft can carry the Senior Span satcom antenna mounted in a large rear-mounted teardrop radome.

Typical missions often reach 10 hours in duration with a standard racetrack profile flown at altitudes around 22860 m (75,000 ft). All U-2s serve with the 9th Reconnaissance Wing headquartered at Beale AFB, CA, with three theatre detachments covering the Mediterranean, Far East and Europe. In 1992, re-engining of U-2Rs began with the GE F101 turbofan, offering greater thrust, endurance, and supportability.

Re-engining with the F101 turbofan (derived from the B-2's F118 engine) will confer a range increase of 15 per cent and restore operational ceiling to above 24340 m (80,000 ft).

Specification: Lockheed TR-1A (now U-2R)
Powerplant: one Pratt & Whitney J75-P-13B non-afterburning turbofan rated at 75.62 kN (17,000 lb st)
Dimensions: wing span 31.39 m (103 ft 0 in); length 19.13 m (62 ft 9 in); height 4.88 m (16 ft 0 in); wing area about 92.90 m² (1,000.00 sq ft)
Weights: basic empty without powerplant and equipment pods less than 4536 kg (10,000 lb); operating empty about 7031 kg (15,500 lb); maximum take-off 18733 kg (41,300 lb); sensor weight 1361 kg (3,000 lb)
Performance: never exceed speed Mach 0.8; maximum cruising speed at 21335 m (70,000 ft) more than 373 kt (692 km/h; 430 mph); maximum rate of climb at sea level about 1525 m (5,000 ft) per minute; climb to 19810 m (65,000 ft) in 35 minutes; operational ceiling 27430 m (90,000 ft); take-off run about 198 m (650 ft) at maximum take-off weight; landing run about 762 m (2,500 ft) at maximum landing weight; maximum range more than 2,605 nm (4828 km; 3,000 miles); maximum endurance 12 hours

Lockheed/Boeing F-22

Production F-22s will retain the YF-22's overall configuration, but will have revised sweep angles for most flying surfaces, smaller vertical fins, further aft intakes and a reprofiled nose.

Lockheed YF-22

The **Lockheed F-22** is the USAF's F-15 replacement and is intended to be the leading American air-to-air fighter at the turn of the century. It meets the USAF's ATF (Advanced Tactical Fighter) requirement for long-range supersonic cruise without afterburning and makes use of low-observ-ables (LO), or stealth technology, to defeat advanced radar defence systems. Boeing and General Dynamics teamed with Lockheed to develop a **YF-22A** candidate for the ATF competition's demonstration/validation tests. The first YF-22A, powered by GE YF120 engines, made its first flight on 29 September 1990. In April 1991, the USAF announced its choice of the F-22 for the ATF production contract.

Configuration

The angular F-22 has a comparatively large, diamond-shaped wing, splayed twin vertical tails and large horizontal tails. The wing blends into the fuselage to provide a lifting body area. Lateral engine air intakes are located in a short, tapered nose. The inlet ducts curve inward and upward, shielding the front faces of the engine from direct illumina-tion by radar. Radar-absorbent materials are employed in the forward fuselage and cockpit canopy. Coupled with large leading-edge wing flaps and overall low wing loading, the F-22's 2-D thrust vectoring engine nozzles permit manoeuvring at low speeds and high flight angles.

Supercruise

The F-22 is intended to cruise at supersonic speed (super-cruise) to a high-risk area and engage opposing aircraft BVR but, if necessary, to be able to outmanoeuvre them at clos-er range.

Engineering and manufacturing development work will be carried out over 1992-96. Production F-22s will differ from the YF-22 introducing revised flying surfaces and sweep angles a blunter nose and shorter intakes. In the FSD F-22 programme, the manufacturers will produce nine single-seat **F-22A** aircraft, two two-seat **F-22B** aircraft, and two ground test vehicles for a total of 13 F-22s.

Lockheed pursued an ambitious test schedule for the YF-22, achieving a first for the ATF programme by firing an AIM-9M missile over the US Navy ranges in December 1990.

Specification: Lockheed/Boeing/General Dynamics F-22A (provisional)
Powerplant: two Pratt & Whitney F119-P-100 turbofans each rated at 155.69 kN (35,000 lb st) with afterburning
Dimensions: wing span 13.56 m (44 ft 6 in); length 18.92 m (62 ft 1 in); height 5.00 m (16 ft 5 in)
Weights: maximum take-off 27216 kg (60,000 lb)
Performance: maximum level speed 'clean' at optimum altitude Mach 1.58 (supercruise mode) and at 9145 m (30,000 ft) Mach 1.7 (afterburning mode); service ceiling 15240 m (50,000 ft)
Armament: one internal M61A2 20-mm cannon, three internal weapons bays, underside bay for four AIM-120A AMRAAMs and two lateral intake bays each with two AIM-9M Sidewinder AAMs; revised bays for 454-kg (1,000-lb) Joint Direct Attack Missile (two JDAMS replacing two AIM-120s) and AIM-9X AAMS; four underwing stores stations with provision for two AGM-137A Tri-Service Standoff Attack Missiles and/or fuel tanks

McDonnell Douglas A-4 Skyhawk

Singapore is a major Skyhawk operator, with two upgraded re-engined A-4S variants in service. Avionics improvements include a Pave Penny laser designator, pilot's HUD and an INS.

The **Douglas A-4 Skyhawk** first flew in prototype form on 22 June 1954, and entered service in October 1956. It provided the US Navy and the USMC with their principal light attack platform for over 20 years. Total production of all variants reached 2,960. Early models comprised J65-powered **A-4A/B/Cs** (differing in avionics and engine power), J52-engined **A-4Es**, and **A-4Fs** with a dorsal avionics hump. Export models comprised **A-4G** (Australia), **A-4H** (Israel), **A-4K** (New Zealand) and **A-4KU** (Kuwait). The **A-4L** was a rebuilt A-4C for the USN reserve. **A-4Ns** were similar to the A-4H but featured uprated avionics, including a HUD. The last major production model was the **A-4M**, based on the A-4F but introducing a J52-P-408A engine.

Few single-seat A-4s are left with the USN, these mostly serving with aggressor squadrons. The major variant is the **A-4F 'Super Fox'**, which has been stripped of most of its attack avionics, including removal of the dorsal hump. The A-4M model remains in limited use with USMC Reserve squadrons. The **'Super Mike'** was a stripped A-4M aggressor model serving with the Naval Air Reserve.

A-4 trainer variants feature two cockpits in tandem with a single canopy, and some combat capability. The USMC's **OA-4M** was used for FAC duties, and has been retired. The definitive **TA-4J** was a simplified version which lacks cannon armament and combat capability. In USN service this is the major operational model, used for advanced training, including carrier qualification.

Foreign operators and foreign upgrades

Current export operators comprise Argentina, Indonesia, Israel and Kuwait. Malaysia's **A-4PTMs** (ex-USN A-4C/Ls) have AGM-65 and AIM-9 capability. Singapore has the most capable Skyhawks, having upgraded surplus A-4B/Cs as **A-4Ss**. Three modification standards exist: the basic refurbished aircraft, the **A-4S-1 Super Skyhawk**, re-engined with the GE F404 turbofan, and the **A-4SU** with F404 and a new digital avionics suite. New Zealand's A-4s have been upgraded with new avionics under Project Kahu.

Project Kahu upgrades 17 A-4K and five TA-4K trainers (shown here) with APG-66 radar and other improved avionics. They can carry AIM-9L AAMS, Maverick AGMs and GBU-16 LGBs.

McDonnell Douglas TA-4J (A-4M side view)

Specification: Singapore Aerospace A-4S-1 Super Skyhawk
Powerplant: one General Electric F404-GE-100D non-afterburning turbofan rated at 48.04 kN (10,800 lb st)
Dimensions: wing span 8.38 m (27 ft 6 in); length 12.72 m (41 ft 8.5 in) including IFR probe; height 4.57 m (14 ft 11.875 in); wing area 24.14 m² (259.82 sq ft)
Weights: operating empty 4649 kg (10,250 lb); maximum take-off 10206 kg (22,500 lb)
Performance: maximum level speed 'clean' at sea level 609 kt (1128 km/h; 701 mph); maximum rate of climb at sea level 3326 m (10,913 ft) per minute; service ceiling 12190 m (40,000 ft); range 625 nm (1158 km; 720 miles) with maximum ordnance
Armament: two Mk 12 20-mm cannon in wingroots; plus ordnance including bombs, rockets, ASMs, AIM-9P AAMs and fuel tanks

McDonnell Douglas KC-10 Extender

Strategic tanker/transport

The KC-10 undertakes dual tanker/transport tasks with both ACC and AMC units. It is also operated by Air Force Reserve crews under the so-called 'associate' programme.

The **McDonnell Douglas KC-10A Extender** strategic tanker/transport is based on the **DC-10 Series 30CF** commercial freighter/airliner and was obtained to satisfy the USAF's ATCA (Advanced Tanker Cargo Aircraft) requirement. Sixteen examples were initially ordered in 1977; procurement was later increased to 60 aircraft. The first Extender made its maiden flight on 12 July 1980 and deliveries to SAC took place between March 1981 and November 1988.

Mission equipment

Changes from commercial DC-10 standard include provision of an IFR receptacle above the cockpit, an improved cargo handling system and some military avionics. A McDonnell Douglas Advanced Aerial Refueling Boom (AARB) is fitted beneath the aft fuselage. The digital FBW control boom can transfer fuel at a rate of 5678 litres (1,249 Imp gal) per minute. The KC-10 is also fitted with a hose and reel unit in the starboard aft fuselage and can thus refuel Navy and USMC aircraft during the same mission. This is a unique capability and one that makes it much more versatile than the KC-135. More recently, wing-mounted HDU pods are to be fitted to all KC-10s so that three receiver aircraft may be refuelled simultaneously with this system.

Fuel capacity

The wing and fuselage fuel cells contain approximately 68610 litres (15,092 Imp gal) and are interconnected with the aircraft's basic fuel system. The KC-10 is able to transfer 90718 kg (200,000 lb) of fuel to a receiver 1,910 nm (3540 km; 2,200 miles) from its home base and return to base. For conventional strategic transport missions the KC-10 has a port-side cargo door and carries standard USAF pallets, bulk cargo or wheeled vehicles. Dual tanker/transport missions include accompanying deploying fighters; this is achieved during the transit by the provision of IFR support. Two ex-Martinair DC-10-30CFs were procured by the Netherlands for conversion by McDonnell Douglas to tanker configuration, and will return to service in 1995.

No fewer than 46 out of 59 USAF KC-10s were deployed for Desert Storm. They flew about 20 per cent of the 4,967 sorties which refuelled aircraft heading into combat.

McDonnell Douglas KC-10A Extender

Specification: McDonnell Douglas KC-10A Extender
Powerplant: three General Electric CF6-50C2 turbofans each rated at 233.53 kN (52,500 lb st)
Dimensions: wing span 47.34 m (155 ft 4 in); length 55.35 m (181 ft 7 in); height 17.70 m (58 ft 1 in); wing area 358.69 m² (3,861.00 sq ft)
Weights: operating empty 108891 kg (240,065 lb); maximum take-off 267620 kg (590,000 lb); aircraft basic fuel system 108062 kg (238,236 lb); fuselage bladder fuel cells 53446 kg (117,829 lb); total internal fuel 161508 kg (356,065 lb); maximum payload 76843 kg (169,409 lb) of cargo
Performance: maximum level speed at 7620 m (25,000 ft) 530 kt (982 km/h; 610 mph); maximum cruising speed at 9145 m (30,000 ft) 490 kt (908 km/h; 564 mph); maximum rate of climb at sea level 884 m (2,900 ft) per minute; take-off distance 3170 m (10,400 ft) at MTOW; landing distance 1868 m (6,130 ft) at maximum landing weight; maximum range with maximum cargo 3,797 nm (7032 km; 4,370 miles)

McDonnell Douglas C-17 Globemaster III   *Long-range heavy transport*

The C-17 will revitalise the US strategic airlift effort after lengthy and costly development. The advanced, conventional design has features such as winglets and FBW controls.

In August 1981 McDonnell Douglas was chosen to proceed with a design to fulfil the USAF's **C-X** requirement for a new heavy cargo transport. The requirement called for the provision of intra-theatre and theatre airlift of outsize loads, including M1 MBTs, armoured vehicles and helicopters, while retaining the ability for tactical delivery profiles, (including LAPES and short landings into austere strips).

Advanced technology
The resulting **C-17A**'s design adopts a classic military transport configuration with a high-mounted supercritical wing, a rear-fuselage loading ramp and undercarriage housings on each side of the fuselage. However, it incorporates such advanced features as winglets, high-performance turbofans (military versions of the Boeing 757's PW2040) and an all-digital FBW control system. The two-crew cockpit is equipped with HUDs and four MFDs. Short-field performance is aided to some extent by an externally-blown flap system based on that demonstrated on the McDonnell Douglas YC-15 prototype.

Typical payloads
The C-17 can be configured for cargo, paratroops, combat troops, hospital litter patients, or combinations of all these. For strategic airlift, it can carry 202 personnel, or 18 USAF 463L pallets. Typical loads include two M2 Bradley AFVs, two Jeeps with trailers, and two 5-ton 8x8 trucks with trailers; three AH-64A Apache helicopters; three AH-1 and four OH-58 helicopters; or air-droppable platforms of up to 49895 kg (110,000 lb). The internal loading system is fully mechanised for one-loadmaster operation.

The C-17 has suffered a protracted devlopment programme. After an earlier FSD schedule had been abandoned, the single prototype (T-1) of the C-17A flew in September 1991, followed by the first three production examples in 1992. The programme received a setback with structural failure of a 'static test' wing. In spite of this, the USAF's 17th Airlift Squadron received its first C-17 in 1993. The USAF intends to procure 120 C-17s, but cutbacks in procurement are likely.

Central to the C-17's STOL capability are its flaps, which are blown with engine exhaust to produce a vectored jet stream.

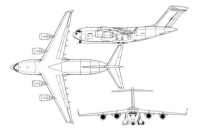

McDonnell Douglas C-17A Globemaster III

Specification: McDonnell Douglas C-17A
Powerplant: four Pratt & Whitney F117-P-100 turbofans each rated at 185.49 kN (41,700 lb st)
Dimensions: span 50.29 m (165 ft 0 in) basic and 52.20 m (171 ft 3 in) between winglet tips; length 53.04 m (174 ft 0 in); height 16.79 m (55 ft 1 in); wing area 353.02 m² (3,800.00 sq ft)
Weights: operating empty 122016 kg (269,000 lb); maximum take-off 263083 kg (580,000 lb); maximum payload 78108 kg (i72,200 lb); typical payload between 56245 kg (124,000 lb) and 69535 kg (153,300 lb)
Performance: maximum cruising speed at low altitude 350 kt (648 km/h; 403 mph) CAS; airdrop speed at sea level between 115 and 250 kt (213 and 463 km/h; 132 and 288 mph); service ceiling 13715 m (45,000 ft); take-off field length with 75750-kg (167,000-lb) payload 2286 m (7,500 ft); landing field length 914 m (3,000 ft) with thrust reversal; radius with 36786-kg (81,100-lb) payload 500 nm (925 km; 575 miles)

McDonnell Douglas F-4 Phantom II

Israel's Kurnass 2000 F-4E upgrade was first flown in 1987, and adds structural strengthening, updated avionics (including a Norden multi-mode radar) and an IFR probe.

McDonnell Douglas F-4EJ Kai Phantom II

The **McDonnell F-4B Phantom** was originally designed as a shipboard interceptor for the USN and USMC. None remain in use as fighters and even the handful operated by test agencies have now been retired, leaving only **QF-4N/S** drones in service. The USAF's initial **F-4C** variant was followed by the **F-4D** optimised for air-to-ground operations. All US F-4C/Ds have been retired, but F-4Ds remain active in Iran and South Korea. The definitive **F-4E** first flew in June 1967, and introduced a 20-mm internal cannon. It serves with Greece, Iran, Israel, South Korea and Turkey. The F-4E has been withdrawn from US service, but some may be converted as drones. Three major operators have upgraded their surviving aircraft, extending airframe lives, and adding modern radar and improved avionics and self-defence systems. They comprise the Israeli **Kurnass 2000**, the Luftwaffe's AMRAAM-capable **F-4F ICE** (Improved Combat Efficiency) and Japan's **F-4EJ Kai**.

'Wild Weasel'

The **F-4G 'Wild Weasel'** anti-radar variant resulted from the conversion of 116 F-4E airframes, deleting the integral cannon and adding an APR-38 RHAWS. The current APR-47 system is compatible with the AGM-45, AGM-65 and AGM-88 missiles for the SEAD role. After highly successful operations in Desert Storm, the F-4G will most likely remain in service with the USAF for some years.

Photo Phantoms

The reconnaissance-configured **RF-4C** was first flown in May 1964 and was used in Vietnam. A modified nose houses optical cameras, electronic reconnaissance equipment, IR sensors and a mapping/terrain avoidance radar. RF-4Cs remain active with Greece, Iran, Turkey and the US ANG. Israel's armed RFs are equipped with indigenous reconnaissance and avionics equipment. Spain operates updated RF-4Cs. The Luftwaffe's 150 **RF-4E**s (similar to the RF-4C) were retired in 1992; some surplus aircraft were passed to Greece and Turkey. Japan operates 14 upgraded **RF-4EJ Kai**s with new TFR, and improved avionics and recce systems.

Germany's ICE programme will add APG-65 radar to 75 F-4Fs, bestowing a capability to launch BVR AIM-120 missiles.

Specification: McDonnell Douglas F-4E Phantom II
Powerplant: two General Electric J79-GE-17A turbojets each rated at 52.53 kN (11,810 lb st) dry and 79.62 kN (17,900 lb st) with afterburning
Dimensions: wing span 11.71 m (38 ft 5 in); length 19.20 m (63 ft 0 in); height 5.02 m (16 ft 5.5 in); wing area 49.24 m² (530.00 sq ft)
Weights: basic empty 13757 kg (30,328 lb); combat take-off 18818 kg (41,487 lb); maximum take-off 28030 kg (61,795 lb)
Performance: maximum level speed 'clean' at 10975 m (36,000 ft) 1,290 kt (2390 km/h; 1,485 mph); cruising speed at MTOW 496 kt (919 km/h; 571 mph); maximum rate of climb at sea level 18715 m (61,400 ft) per minute; service ceiling 18975 m (62,250 ft); area interception combat radius 683 nm (1266 km; 786 miles)
Armament: one M61 20-mm cannon with 640 rounds, maximum ordnance 7258 kg (16,000 lb) including four AIM-7 and four AIM-9 AAMs, plus a wide variety of bombs, rockets, LGBs, fuel tanks

McDonnell Douglas F-15A/C Eagle

Eagle supreme: the F-15 is a BVR interceptor par excellence, and is also an able dogfighter. The latest MSIP fighter model served with distinction in Desert Storm, accounting for the majority of Iraqi aircraft destroyed by coalition forces.

The **McDonnell Douglas F-15 Eagle** air superiority fighter and interceptor is widely viewed as setting the world standard for the primarily BVR air-to-air mission it performs. It was designed to meet the USAF's 1968 FX requirement which called for a long-range tactical air superiority fighter to replace the F-4. McDonnell won the competition and flew a prototype **F-15A** on 27 July 1972, followed by a prototype **F-15B** two-seat trainer in July 1973.

Radar and systems
The F-15 has an advanced aerodynamic design with large lightly-loaded wings conferring high agility. It features a sophisticated avionics system and its APG-63 radar introduced a genuine look-down/shoot-down capability. Radar-guided AIM-7 AAMs form the primary armament, augmented by AIM-9 AAMs and an internal 20-mm M61A1 cannon for closer quarter combat.

The USAF received 360 production F-15As and 58 F-15Bs from 1976. Most F-15A/Bs serve in training and air superiority duties with ANG units. The only foreign F-15A/B operator is Israel, whose Eagles were the first to see combat, claiming five Syrian MiG-21s in 1979. Over 40 aircraft (mostly Syrian MiGs) were destroyed over the Bekaa Valley in 1982.

F-15C/D
The **F-15C** is the definitive production F-15 fighter model and represents an improved and updated F-15A derivative. The two-seat **F-15D** similarly succeeds the F-15B trainer. First flying on 26 February 1979, the F-15C introduced uprated F100 engines and provision for conformal fuel tanks (CFTs). Initial deliveries were made in September 1979 and F-15C/Ds later replaced F-15A/Bs with three wings.

F-15C/Ds were delivered to the USAF (408/62), Israel (23/8) and Saudi Arabia (55/19). USAF and Saudi F-15Cs were heavily involved in Desert Storm, scoring 32 aerial victories with no losses. The equivalent **F-15J/DJ** forms Japan's principal air superiority type. Most of the JASDF's 191 planned Eagles will have been assembled by Mitsubishi.

The 5th FIS had a brief life in F-15As until being disbanded in 1988. It was tasked with defending the northern United States.

McDonnell Douglas F-15C (F-15D side view)

Specification: McDonnell Douglas F-15C Eagle
Powerplant: two Pratt & Whitney F100-P-220 turbofans each rated at 65.26 kN (14,670 lb st) dry and 106.0 kN (23,830 lb st) with afterburning
Dimensions: wing span 13.05 m (42 ft 10 in); length 19.43 m (63 ft 9 in); height 5.63 m (18 ft 5.5 in); wing area 56.48 m² (608.00 sq ft)
Weights: operating empty 12793 kg (28,600 lb); normal take-off 20244 kg (44,630 lb) on an interception mission with four AIM-7 AAMs; maximum take-off 30844 kg (68,000 lb) with CFTs
Performance: maximum level speed 'clean' at 10975 m (36,000 ft) more than 1,433 kt (2655 km/h; 1,650 mph); maximum rate of climb at sea level more than 15240 m (50,000 ft) per minute; service ceiling 18290 m (60,000 ft); combat radius 1,061 nm (1967 km; 1,222 miles) (interception mission)
Armament: one M61 20-mm cannon with 940 rounds; maximum ordnance 7257 kg (16,000 lb); primary armament comprises four AIM-7M and four AIM-9M IR-homing short-range AAMs

McDonnell Douglas F-15E Eagle

The versatile F-15 airframe served as the basis for further development into a superb all-weather attack platform. The F-15E can launch PGMs and other weapons, but retains the fighter's air-to-air combat capabilities.

The F-15 was originally intended as dual-role aircraft, incorporating air-to-ground capability and wired for the carriage of air-to-ground ordnance. This ground attack role was abandoned in 1975, but later resurrected in 1982, when the second TF-15A was modified as the privately-developed **'Strike Eagle'**. It was conceived as a replacement for the F-111. Development of the the resulting **F-15E** began in February 1984 and the first production aircraft made its maiden flight on 11 December 1986.

The F-15E's primary mission is air-to-ground strike, for which it carries a wide range of weapons on two underwing pylons, underfuselage pylons and 12 bomb racks mounted directly on the CFTs. It introduces redesigned controls, a wide field of vision HUD, and three multi-purpose CRTs displaying navigation, weapons delivery and systems operations. The rear-cockpit WSO employs four multi-purpose CRT terminals for radar, weapon selection and monitoring of enemy tracking systems. The WSO also operates an AN/APG-70 synthetic aperture radar and LANTIRN navigation and targeting pods. The navigation pod incorporates its own TFR, which can be linked to the aircraft's flight control system to allow automatic coupled terrain following flight. The targeting pod allows the aircraft to self-designate LGBs. The F-15E's original F100-PW-220 turbofans were soon replaced by P&W's F100-PW-229 engine under the Improved Performance Engine competitive programme. Operational F-15Es were delivered to the USAF in 1989 and currently equip 10 squadrons.

Exports and research variant

F-15Es are to be exported to Israel and (in downgraded form) Saudi Arabia. The TF-15 was modified as the **SMTD** (STOL/Maneuver Technology Demonstrator), designed to operate from much shortened runways. It was equipped with P&W two-dimensional thrust-vectoring nozzles and canard foreplanes to improve low-speed performance. First flying on 7 September 1988, it demonstrated an ability to land under a simulated 61-m (200-ft) ceiling in total darkness.

This F-15E carries the 2,000-lb AGM-130 PGM, along with LANTIRN pods, AIM-7 missiles and an AXQ-14 datalink pod.

McDonnell Douglas F-15E 'Strike Eagle'

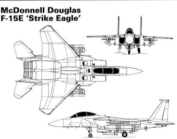

Specification: McDonnell Douglas F-15E Eagle
Powerplant: two F100-PW-229 turbofans each rated at 79.18 kN (17,800 lb st) dry and 129.45 kN (29,100 lb st) with afterburning
Dimensions: wing span 13.05 m (42 ft 10 in); length 19.43 m (63 ft 9 in); height 5.63 m (18 ft 5.5 in); wing area 56.48 m² (608.00 sq ft)
Weights: operating empty 14379 kg (31,700 lb); maximum take-off 36741 kg (81,000 lb)
Performance: maximum level speed 'clean' at high altitude more than 1,433 kt (2655 km/h; 1,650 mph); maximum rate of climb at sea level more than 15240 m (50,000 ft) per minute; combat radius 685 nm (1270 km; 790 miles)
Armament: one M61A1 20-mm cannon with 940 rounds and 4 AIM-9M AAMs for self-defence; maximum ordnance 11000 kg (24,250 lb), including Rockeye or CBU-87 cluster bombs; 'iron' bombs; GBU-10, GBU-12, GBU-15 LGBs; CBU-52, -58, -71, -87, -89, -90, -92 or -93 bombs; AGM-65 AGMs; B57 or B61 nuclear bombs; AIM-7Ms or AIM-120 AMRAAMs; AGM-88 HARMs

McDonnell Douglas F/A-18A/C Hornet

An Australian Hornet strike fighter of No. 2 OCU pulls hard, its LERXes generating the powerful vortices which give the F/A-18 its superlative high-Alpha capability.

Emerging victorious in the USN's Air Combat Fighter programme, the **Hornet** was a more sophisticated navalised derivative of the **Northrop YF-17** contender. A single common **F/A-18** fighter/attack aircraft was eventually developed (replacing USN A-7s and USMC F-4s), with responsibility for development and production shared by McDonnell Douglas and Northrop. The first of 11 pre-production aircraft made the type's maiden flight on 18 November 1978 and production followed of 371 **F/A-18As**.

**McDonnell Douglas
F/A-18A Hornet**

True multi-role capability
The F/A-18 introduced a genuinely multi-role capability. The pilot's well-designed cockpit has three multi-function CRT type displays and true HOTAS controls, enabling him to switch easily from the air-to-ground role to air-to-air or defence suppresion duties. The F/A-18's dogfighting capability is remarkable, advanced wing design with large slotted LERXes conferring excellent high-Alpha capability and turn performance. Similarly, the multi-mode APG-65 radar is as effective at putting bombs with high accuracy on target as it is at detecting and engaging multiple airborne targets.

Improved C model
The improved **F/A-18C** was first flown in September 1986. An expanded weapons capability introduces compatibility with both AIM-120 AMRAAM and imaging IR AGM-65 missiles. The F/A-18C also features an avionics upgrade with new AN/ALR-67 RHAWS, provision for the cancelled AN/ALQ-165 airborne self-protection jammer and improvements to mission computer equipment. After 137 baseline F/A-18Cs had been delivered, production switched to a night-attack capable version, featuring compatability with Cat's Eyes PNVGs, a Hughes AN/AAR-50 TINS (Thermal Imaging Navigation Set) pod presenting its thermal picture of the terrain ahead in the raster HUD, externally-carried AN/AAS-38 targeting FLIR pod and colour MFDs.

F/A-18As were exported to Australia (57 **AF-18s**), Canada (98 **CF-188As**) and Spain (60). F/A-18Cs have been sold to Finland (57), Kuwait (32) and Switzerland (26).

The 'Black Knights' of VMFA-314 were the first front-line F/A-18A unit, and took them into action against Libya in 1986.

Specification: McDonnell Douglas F/A-18C Hornet
Powerplant: two F404-GE-402 turbofans each rated at 78.73 kN (17,700 lb st) with afterburning
Dimensions: wing span 12.31 m (40 ft 5 in) with tip-mounted AAMs; width folded 8.38 m (27 ft 6 in); length 17.07 m (56 ft 0 in); height 4.66 m (15 ft 3.5 in); wing area 37.16 m² (400.00 sq ft)
Weights: empty 10455 kg (23,050 lb); normal take-off 16652 kg (36,710 lb) for a fighter mission or 23541 kg (51,900 lb) for an attack mission
Performance: maximum level speed 'clean' at high altitude more than 1,033 kt (1915 km/h; 1,190 mph); maximum rate of climb at sea level 13715 m (45,000 ft) per minute; combat radius over 400 nm (740 km; 460 miles) on a fighter mission
Armament: one M61A1 20-mm cannon with 570 rounds; maximum ordnance 7031 kg (15,500 lb) including AIM-120, AIM-7 and AIM-9 AAMs; PGMs (AGM-65, AGM-62 and GBU-10/12/16 LGBs), Mk 80 series of bombs, CBU-59 cluster bombs units and fuel air explosives; AGM-84 ASMs; B57 and B61 tactical nuclear devices and AGM-88A HARMs

McDonnell Douglas F/A-18B/D Hornet

Fighter-bomber

The F/A-18D forms the basis of the USMC's all-weather attack squadrons. The VMFA (AW)-533 aircraft in the foreground recorded the two millionth Hornet flight hour.

A two-seater Hornet version was developed concurrently with that of the single-seater. Two **TF-18A** examples were procured initially for RDT&E tasks. The designation was later replaced by **F/A-18B**. Basically identical to the F/A-18A, provision of a second seat in tandem was accomplished at a six per cent penalty in fuel capacity. The F/A-18B is otherwise unaltered, retaining identical equipment and near-identical combat capability. The **F/A-18D** two-seat variant is broadly similar to the single-seat F/A-18C. Thirty-one baseline aircraft were procured before production switched to the night-attack-capable F/A-18D which has the same avionics improvements as the night-attack F/A-18C. The night-attack F/A-18D forms the basis of a special night-attack aircraft now replacing the A-6 with the Marine Corps All Weather Attack Squadrons. Originally dubbed **F/A-18D+**, the aircraft has 'uncoupled' cockpits (no rear seat control column) with two sidestick weapons controllers.

Recce platform
The increasing trend towards the use of digitised reconnaissance sensors led to the development of a new variant for the USMC. The modified **F/A-18D(RC)** version was originally intended to have the same recce nose as the single-seat reconnaissance Hornet, but also uses the ATARS pod which contains a high resolution synthetic aperture SLAR, transmitting imagery in real time by datalink. The old-style recce nose was abandoned and ATARS was redesigned to be able to be packaged into the nose, but delivery of the F/A-18D(RC) to the USMC in February 1992 marked only the delivery of aircraft wired for ATARS.

F/A-18E/F
Originally proposed during 1991 as a replacement for the abandoned A-12 Avenger project, extensively redesigned **F/A-18E/F** single- and two-seat Hornets are presently under development with a view to first flight in 1995. These are fundamentally stretched F/A-18C/Ds, featuring lengthened fuselages, as well as increased span and wing area, enlarged horizontal tails and enlarged LERXes.

The first European F-18 customer was Spain, whose 12 EF-18Bs (locally designated CE.15) are all based at Zaragoza.

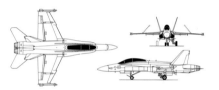

McDonnell Douglas F/A-18B

Specification: McDonnell Douglas F/A-18B Hornet generally similar to the F/A-18A Hornet except in the following particulars
Weights: normal take-off 15234 kg (33,585 lb) for a fighter mission; maximum take-off 21319 kg (47,000 lb) for an attack mission
Range: ferry range with internal and external fuel 1,900 nm (2,187 miles; 3520 km); combat radius 550 nm (1020 km; 634 miles) on an attack mission
Armament: (Australia) AIM-7 and AIM-9 AAMs, plus 2,000-lb LGBs, AGM-84 ASMs and AGM-88 HARMs; (Canada) LAU-5003 pods containing 19 CRV-7 69.85-mm (2.75-in) high-velocity rockets and BL 755 cluster bombs; (Spain) AIM-7F and AIM-9L AAMs, plus a variety of bombs for surface attack, and AGM-88 HARMs and AGM-84 ASMs; (USMC Fast FAC) two four-round LAU-10 launchers for 127-mm (5-in) Zuni rockets, two 1250-litre (273-Imp gal) fuel tanks and two AIM-9L/M AAMs, plus AAR-50 TINS or AAS-38 FLIR in port AIM-7F/M Sparrow fuselage station and an ASQ-173 laser spot tracker/strike camera on starboard station

McDonnell Douglas AV-8B Harrier II   *STOVL close-support*

The Harrier II Plus is the latest and most capable version, with enhanced air-to-air capability. The APG-65 radar endows BVR kill ability with AIM-7 and AMRAAM missiles.

After unsuccessfully pursuing their own programmes for an advanced successor to the Harrier/AV-8A with a larger war-load and improved range/endurance characteristics, BAe and McDonnell Douglas collaborated on the joint **Harrier II** programme. The new design featured a new, larger-area carbon-fibre supercritical wing, a completely revised, raised cockpit, and advanced aerodynamic lift-enhancing devices including LERXes and underfuselage lift-increasing 'dams'. The new wing was first flown on 9 November 1978, fitted to the 11th AV-8A (which became the first of two prototype **YAV-8B**s). The USMC took delivery of the first production **AV-8B**s in 1983 and used the type during Desert Storm, during which it was heavily engaged on ground support strikes. From the 167th airframe on, all USMC AV-8Bs were made night-attack capable with the installation of a FLIR, a head-down display, a colour moving map and an improved HUD. The terms **Night Attack Harrier II** or **Night Attack AV-8B** are sometimes applied unofficially.

For training, McDonnell Douglas developed the **TAV-8B** with a new forward fuselage, accommodating stepped tandem cockpits. Internal fuel is reduced by 205 kg (453 lb). To offset the reduced stability caused by the longer fuselage, the vertical fin is increased in area by adding 0.43 m (1 ft 5 in) to the height and widening the chord by straightening the lower portion of the leading edge.

Improved radar-equipped Harrier II Plus

The 205th production single-seater was the first fully equipped example of the improved **AV-8B Harrier II Plus** variant. Equipped with the Hughes AN/APG-65 radar, the Harrier II Plus retains the overnose FLIR sensor and is otherwise externally identical to late production AV-8Bs. The provision of radar gives compatability with the AIM-7 Sparrow and AIM-120 AMRAAM (endowing a BVR-kill capability), and AGM-84 Harpoon missiles for anti-shipping duties. Two other nations have also purchased the AV-8B, these being Spain (one TAV-8B, 12 **EAV-8B**s and 13 Harrier II Plus) and Italy (two TAV-8B and 16 Harrier II Plus).

In US Marine Corps service, the AV-8B provides the bulk of close air support for amphibious and helicopter assaults. Five AV-8B squadrons were committed to Desert Storm.

McDonnell Douglas AV-8B (TAV-8B side view)

Specification: McDonnell Douglas AV-8B Harrier II (aircraft delivered after December 1990)
Powerplant: one Rolls-Royce F402-RR-408 turbofan rated at 105.87 kN (23,800 lb st)
Dimensions: wing span 9.25 m (30 ft 4 in); length 14.12 m (46 ft 4 in); 3.55 m height (11 ft 8 in); wing area 22.61 m^2 (343.40 sq ft)
Weights: operating empty 6336 kg (13,968 lb) ; normal take-off 10410 kg (22,950 lb); maximum take-off 14061 kg (31,000 lb) for 405-m (1,330-ft) STO or 8596 kg (18,950 lb) for VTO
Performance: maximum level speed 575 kt (1065 km/h; 662 mph); maximum rate of climb 4485 m (14,715 ft) per minute; combat radius 90 nm (167 km; 103 miles)
Armament: one GAU-12A 25-mm cannon (optional) with 300 rounds; maximum ordnance 6003 kg (13,235 lb), including AGM-65s, Rockeye cluster bombs and AIM-9Ls for self-defence

McDonnell Douglas T-45 Goshawk   *Land-based carrier trainer*

Basically similar to the standard Hawk trainer, the T-45 is significantly strengthened to cope with carrier operations. Initial deliveries were made to VT-21 at Kingsville, Texas.

In 1981, a modified **British Aerospace Hawk** was selected by the US Navy as its **T45TS** (Training System) with McDonnell Douglas as prime contractor. After lengthy evaluation the Hawk was judged superior to existing Navy trainers and rivals, and in November 1981 was duly selected. The principal subcontractor is British Aerospace, which constructs wings, centre and rear fuselage, fin, tailplane, windscreen, canopy and flying controls. As first proposed there were to be two variants, a 'wet' **T-45A** model fitted for carrier operation and a 'dry' model **T-45B** restricted to land-based training and dummy carrier landing practice. Life extension of the T-2 and TA-4J led to a decision to acquire only the T-45A.

US Navy modifications

In order to tailor the basic Hawk airframe to meet stringent US Navy requirements for carrier operation, the aircraft has a strengthened twin nosegear, compatible with the US Navy's steam catapults. The main gear is redesigned, with longer stroke oleos. Fin height and tailplane span are increased and a single ventral fin is added. The ventral airbrake is replaced by two fuselage side-mounted units. The T-45 has new full-span leading-edge slats and is provided with an arrester hook, US Navy standard cockpit instrumentation and radios, Martin-Baker Mk 14 NACES ejection seats and a revised fuel system.

Although strengthened for carrier operation, the T-45A (named **Goshawk** to avoid confusion with the US Army Hawk missile) will remain land-based, flying to a training carrier as required. Two F405-RR-400L (Adour 861-49) engined FSD prototypes were followed by F405-RR-401 (Adour 871) engined pre-production aircraft. The 268 aircraft required (plus 24 simulators, and 34 computer-aided instruction devices) are being delivered to NAS Kingsville, NAS Chase Field and NAS Meridian. A glass cockpit with two colour MFDs will be added from the 97th aircraft. Introduction of the T-45 will result in the training task being accomplished with a reduction of 25 per cent in flying hours, 42 per cent in aircraft and 46 per cent in personnel.

This T-45 captures the arresting wire in its historic first carrier landing aboard USS Kennedy on 4 December 1991.

McDonnell Douglas T-45A Goshawk

Specification: McDonnell Douglas/British Aerospace T-45A Goshawk
Powerplant: one Rolls-Royce/Turboméca F405-RR-401 (Adour 871) turbofan rated at 26.00 kN (5,845 lb st)
Dimensions: wing span 9.39 m (30 ft 10 in); length 11.97 m (39 ft 3 in) including probe; height 4.27 m (14 ft 0 in); wing area 16.69 m² (176.90 sq ft)
Weights: empty 4263 kg (9,399 lb); maximum take-off 5787 kg (12,758 lb)
Performance: maximum level speed 'clean' at 2440 m (8,000 ft) 538 kt (997 km/h; 620 mph); maximum rate of climb at sea level 2128 m (6,982 ft) per minute; service ceiling 12875 m (42,250 ft); take-off distance to 15 m (50 ft) 1189 m (3,744 ft) at maximum take-off weight; landing distance from 15 m (50 ft) 1189 m (3,900 ft) at maximum landing weight; ferry range on internal fuel 1,000 nm (1854 km; 1,152 miles)

McDonnell Douglas AH-64 Apache

The Apache carries Hellfire missiles for precision attack against armour or bunkers, or rocket pods for area suppression. It has sophisticated sensors for search and acquistion of targets.

McDonnell Douglas AH-64D Longbow Apache

Hughes' **AH-64A Apache** was developed to meet a US Army requirement for an advanced attack helicopter (AAH) suitable for the day/night/adverse weather anti-armour role. The AH-64 is a tandem, two-seat helicopter with armoured structure, advanced crew protection systems, avionics, electro-optics, and weapon-control systems, including a TADS/PNVS (Target Acquisition and Designation Sight/Pilot's Night Vision Sensor).

Operations and exports

The Army intends to procure 813 AH-64s. The first combat unit was declared operational in July 1986 and the ARNG became an operator in 1989. The first combat deployment took place in Panama in late 1989. More recently, Apaches were employed to devastating effect during Desert Storm. On 17 January 1992, Apaches fired the first shots of the war, attacking Iraqi radar sites. Despite its high price and indifferent maintenance/reliability record, the Apache has been exported to Egypt, Israel and Saudi Arabia. It has been ordered by Greece and the United Arab Emirates.

Upgraded models – Longbow Apache

The US Army hopes to upgrade 254 AH-64As to interim **AH-64A+** or **AH-64B** standard with near-term improvements based on Desert Storm experience. Modifications include addition of GPS, extended-range fuel tanks, new rotor blades and laser warning receivers and in-field retrofitted GE T700-GE-701C engines. In 1995 the **AH-64C** will enter service, but will lack the Long-bow radar/missile system fitted to the more advanced **AH-64D Longbow Apache** which is due to follow in 1997. This will have a mast-mounted Longbow radar, T700-GE-701C engines and new RF seeker-equipped fire-and-forget Hellfire ATGMs. The AH-64D will also incorporate a range of improvements in targeting, battle management, communications, weapons and navigation systems, plus new glass cockpit displays and symbology. The US Army plans to convert 535 existing AH-64As (308 AH-64Cs, 227 AH-64Ds). One potential AH-64D customer is the British Army.

The AH-64 can designate for its own laser-guided missiles, or work with an airborne scout in a light helicopter.

Specification: McDonnell Douglas Helicopters AH-64A Apache (from 604th helicopter)
Powerplant: two General Electric T700-GE-701C turboshafts each rated at 1342 kW (1,800 shp)
Dimensions: main rotor diameter 14.63 m (48 ft 0 in); wing span 5.23 m (17 ft 2 in) clean; length overall, rotors turning 17.76 m (58 ft 3 in) and fuselage 14.97 m (49 ft 1.5 in); height overall 4.66 m (15 ft 3.5 in) to top of air data sensor; main rotor disc area 168.11 m² (1,809.56 sq ft)
Weights: empty 5165 kg (11,387 lb); normal take-off 6552 kg (14,445 lb) at primary mission weight; maximum take-off 9525 kg (21,000 lb)
Performance: maximum level speed 'clean' 158 kt (293 km/h; 182 mph); maximum vertical rate of climb at sea level 762 m (2,500 ft) per minute; range 260 nm (428 km; 300 miles) with internal fuel
Armament: one M230 Chain Gun 30-mm cannon with 1,200 rounds; (principal armament) up to 16 AGM-114A Hellfire long-range, laser-guided ATGMs; other ordnance includes 19-shot 70-mm (2.75-in) Hydra 70 rocket pods; optional AIM-9L, AIM-92A Stinger and Mistral AAMs

McDonnell Douglas MD-500

The Kenyan air force operates nearly 50 MD-500s on a variety of duties. This is one of 15 MD-500/TOW anti-armour Defenders, equipped with a nose-mounted sight.

Hughes' **YHO-6** design was developed to meet a 1960 US Army requirement for a light observation helicopter. A prototype flew on 27 February 1963 and featured an egg-shaped cabin and an innovative four-bladed rotor, endowing excellent manoeuvrability. The production **OH-6A Cayuse** entered service in 1965 and was widely used in Vietnam. Many surviving 'Loaches' (mostly in Reserve units) are to be modified to **OH-6B** standard with a T63-A-720 engine, and an undernose FLIR.

The civilian **Hughes 500** introduced an uprated engine, increased fuel and a revised interior. The first military variant was the **Model 500M Defender**. The **500M/ASW** has a MAD bird and can carry torpedoes. The civilian **500D** variant introduced a slow-turning five-bladed rotor and a T-tail. It was built under licence in Japan as the **OH-6D**. The basic military 500D is the **500MD Defender** with armour protection and IR exhaust suppression. Variants have been developed for ASW, anti-tank and armed scout duties. Army Special Operations models comprise **EH-6E** (Elint/Sigint/Command Post), **MH-6E** (insertion) and **AH-6F** (attack) aircraft. The AH-6F has a mast mounted sight, an M230 Chain Gun, and can carry pairs of Stinger AAMs.

Next-generation models

The **Model 500E** introduced a revised, pointed nose, more spacious interior and an Allison 250-C20B engine. Dedicated military models are designated **500MG Defender**. The civilian-aimed **MD530F Lifter** introduces a larger diameter main and tail rotors and an Allison 250-C30 turboshaft. The military **MD530MG Defender** has options for a mast-mounted TOW sight, FLIR, RHAW gear, IFF and a laser rangefinder, and can be armed with Hughes TOW 2 missiles, 2.75 in rockets, Stinger AAMs and an M230 Chain Gun. The US Army's 'Little Birds' are reportedly based on the non-NOTAR)no tail rotor) MD530FF, but have the original rounded nose contours of earlier MD500s. These comprise the **MH-6H** (transport/insertion) and **AH-6G** (gunship) helicopters.

This Defender's mast-mounted sight allows it to scout for targets with greater safety, popping out from cover only to fire TOW anti-armour missiles.

McDonnell Douglas MD-530MG Defender

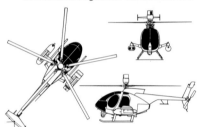

Specification: McDonnell Douglas Helicopters (Hughes) Model 500
Powerplant: one 236-kW (317-shp) Allison 250-C18A turboshaft derated to 207 kW (278 shp) for take-off and 181 kW (243 shp) for continuous running
Dimensions: main rotor diameter 8.03 m (26 ft 4 in); length overall, rotors turning 9.24 m (30 ft 3.75 in) and fuselage 7.01 m (23 ft 0 in); height 2.48 m (8 ft 1.5 in) to top of rotor head; main rotor disc area 50.60 m² (544.63 sq ft)
Weights: empty 493 kg (1,088 lb); normal take-off 1157 kg (2,550 lb); maximum take-off 1361 kg (3,000 lb)
Performance: maximum level speed at 305 m (1,000 ft) 132 kt (244 km/h; 152 mph); maximum rate of climb at sea level 518 m (700 ft) per minute; service ceiling 4390 m (14,400 ft); hovering ceiling 2500 m (8,200 ft) in ground effect and 1615 m (5,300 ft) out of ground effect; range 267 nm (606 km; 307 miles)

Mikoyan MiG-21 'Fishbed'

MiG-21s are still in widespread service, and there are several upgrade programmes available. This Russian aircraft is seen with modern R-27 'Alamo' and R-73 'Archer' missiles.

Mikoyan-Gurevich MiG-21'Fishbed'

The original **MiG-21** was developed as a lightly armed and simply equipped interceptor. Early model **MiG-21F-13 'Fishbed-Cs'** were armed with one NR-30 cannon, and two AA-2 'Atoll' AAMs (or rocket pods), while the **MiG-21P 'Fishbed-D'** dispensed with the cannon armament altogether, but introduced R1L radar. The R-11F2-300-engined **MiG-21PF** was externally indistinguishable, although late **'Fishbed-Es'** introduced a broader-chord fin and had provision for an external GP9 cannon pod. The R-11F-300-engined **MiG-21FL** was for export. The **MiG-21PFS** and **MiG-21PFM** had two-piece canopies, blown SPS flaps and the R-11F2S-300 engine. Small numbers of early 'Fishbeds' remain in service. India has three squadrons of MiG-21FLs.

Later variants

All later variants had blown flaps, two-piece canopies, broad-chord tailfins and four pylons. The MiG-21PFM-based reconnaissance **MiG-21R** had an enlarged dorsal fairing and provision for centreline reconnaissance pods. The **MiG-21S** was similar, with a centreline GP9. The R-13-300-engined **MiG-21SM** put the GSh-23L cannon in a fixed installation, instead of in the removeable GP9 gondola. The R-11F2S-300 engined export **MiG-21M** was also built under licence in India, while the **MiG-21MF** introduced AAM capability on all four pylons. The **MiG-21MT** used the more powerful R-13F-300 engine, while the **MiG-21SMT 'Fishbed-K'** was fitted with a further enlarged spine. Large numbers of MiG-21Rs, MiG-21Ms and MiG-21MFs remain in service. Many of the proposed retrofit programmes currently would be applied to these variants. The multi-role R-25-300-powered **MiG-21bis** introduced improved avionics, AA-8 'Aphid' AAMs, a redesigned spine and improved Sapphire-21 radar. It remains in widespread service.

The unarmed **MiG-21U 'Mongol-A'** tandem two-seat trainer could carry a centreline gun pod and had two underwing pylons. The **MiG-21US 'Mongol-B'** had increased fin chord, improved ejection seats, a bigger spine, a retractable periscope and blown SPS flaps. The **MiG-21UM** was similar, with updated instruments and avionics.

Some 75-100 Egyptian MiG-21s have been upgraded with Western avionics. They can carry AIM-9P-3 and R550 AAMs.

Specification: Mikoyan-Gurevich MiG-21bis 'Fishbed-M'
Powerplant: one Tumanskii R-25-300 turbojet rated at 69.58 kN (15,650 lb st) with afterburning
Dimensions: wing span 7.15 m (23 ft 5.5 in); length 15.76 m (51 ft 8.5 in) including probe; height 4.12 m (13 ft 6.2 in); wing area 22.95 m^2 (247.03 sq ft)
Weights: empty 5350 kg (11,795 lb); normal take-off 8212 kg (18,104 lb) with four AAMs; maximum take-off 9661 kg (21,299 lb)
Performance: maximum level speed 'clean' at 11000 m (36,090 ft) 1,203 kt (2230 km/h; 1,385 mph); maximum level speed at sea level 621 kt (1150 km/h; 715 mph); maximum rate of climb at sea level 7200 m (23,622 ft) per minute; service ceiling 19000 m (62,336 ft); typical combat radius 243-270 nm (450-500 km; 280-311 miles)
Armament: one centreline twin-barrelled GSh-23 23-mm cannon, maximum ordnance 2000 kg (4,409 lb), including four UV-16-57 rocket pods, or four 240-mm rockets, or two 500-kg (1,102-lb) and two 250-kg (551-lb) bombs, or four AAMs

Mikoyan MiG-23 'Flogger'

Intercepted by US Navy fighters, this Libyan 'Flogger-E' carries four AA-2 'Atoll' missiles. This downgraded variant has 'Jay Bird' radar in a short radome, and no BVR missile capability.

The **MiG-23** was developed as a STOL MiG-21 replacement, with greater range and firepower. The delta-winged Model 23-01 prototype had lift jets and was evaluated against the VG **Model 23-11**. Powered by a Tumanskii R27F-300 turbojet, the 23-11 first flew on 10 April 1967 and was ordered into production as the **MiG-23S**. The **MiG-23M 'Flogger-B'** had pulse-Doppler radar, an IRST and AA-7 'Apex' SARH missiles. Two export versions were the **MiG-23MS 'Flogger-E'** with 'Jay Bird' radar, and no BVR missile, while the **MiG-23MF 'Flogger-B'** retained 'High Lark' radar, and AA-7 missiles.

The lightweight **MiG-23ML 'Flogger-G'** introduced airframe, engine, radar and avionics improvements and was delivered to Frontal Aviation, North Korea, Czechoslovakia and East Germany. The similar **MiG-23P** for the PVO could be automatically steered onto its target by GCI. The ultimate **MiG-23MLD 'Flogger-K'** fighter variant incorporates vortex generators on the pitot probe and notches in its vestigial LERXes to improve high Alpha handling.

Attack variants

Mikoyan began studies of a new jet *shturmovik* during 1969, before economic constraints forced them to develop a MiG-23S-based derivative. The **MiG-23B** introduced a more sloping radarless nose for improved view, and a 112.78-kN (25,353-lb) Lyul'ka AL-21F-300 engine in a shortened rear fuselage. Production subsequently switched to the **MiG-23BN**, which featured an upgraded nav/attack system, and was powered by a derated R-29B-300 engine. Both variants share the reporting name **'Flogger-F'**.

Two new variants introduced improved avionics, both known as **'Flogger-H'**. The **MiG-23BK** had the same nav/attack system and laser rangefinder as the MiG-27K. The **MiG-23BM** was similar, with the nav/attack system of the MiG-27D. Attack 'Floggers' were exported to Bulgaria, Cuba, Czecoslovakia, East Germany, India and Iraq.

The **MiG-23UB** tandem two-seat trainer version serves with all MiG-23/-27 operators. It features a retractable instructor's periscope to give better view forward on approach.

The MiG-23ML 'Flogger-G' has a lightened airframe, an improved SP-23L radar, an R-35-300 engine and a new IRST.

Mikoyan-Gurevich MiG-23MF 'Flogger-B'

Specification: Mikoyan-Gurevich MiG-23ML 'Flogger-G'
Powerplant: one MNPK 'Soyuz' (Khachatourov) R-35-300 turbojet rated at 83.84 kN (18,849 lb st) dry and 127.49 kN (28,660 lb st) with afterburning
Dimensions: wing span 13.97 m (45 ft 9.8 in) spread and 7.78 m (25 ft 6.25 in) swept; length 16.70 m (54 ft 9.5 in); height 4.82 m (15 ft 9.75 in); wing area 37.35 m² (402.05 sq ft) spread and 34.16 m² (367.71 sq ft) swept
Weights: empty 10200 kg (22,487 lb); normal take-off 14700 kg (32,407 lb); maximum take-off 17800 kg (39,242 lb)
Performance: maximum level speed 'clean' at 11000 m (36,090 ft) 1,349 kt (2500 km/h; 1,553 mph); maximum rate of climb at sea level 14400 m (47,244 ft) per minute; service ceiling 18500 m (60,695 ft); combat radius 620 nm (1150 km; 715 miles) with six AAMs
Armament: one twin-barrelled GSh-23 23-mm cannon; maximum ordnance 3000 kg (6,614 lb), (intercept) up to two AA-7 and six AA-8 AAMs

88

Mikoyan MiG-25 'Foxbat'

The 'Foxbat-B' codename covers several tactical reconnaissance variants with a mix of cameras and radar in the nose.

The **MiG-25** was developed to counter the Mach 3 XB-70 Valkyrie strategic bomber. It featured advanced construction techniques, using tempered steel for most of the airframe and titanium for the nose and leading edges. The prototype **Ye-155P-1** flew on 9 September 1964, powered by a pair of 100-kN (22,500-lb) Mikulin R-15B-300 turbojets.

Production of the refined **MiG-25P 'Foxbat-A'** fighter began in 1969, but the aircraft did not enter service until 1973. The definitive **MiG-25PD 'Foxbat-E'** interceptor featured a new RP-25 look-down/shoot-down radar, an IRST, more powerful R-15BD-300 turbojets and provision for a large 5300-litre (1,166-Imp gal) belly tank. Surviving 'Foxbat-As' were brought up to PD standard, as the **MiG-25PDS**, often with a 250-mm (10-in) nose plug to allow installation of a retractable IFR probe. The **MiG-25PU 'Foxbat-C'** conversion trainer lacks radar and has a new instructor's cockpit stepped down in front of the standard cockpit. MiG-25 fighters were exported to Algeria, Iraq, Libya and Syria.

Reconnaissance models
The prototype **Ye-155R-1** flew six months before the prototype fighter (on 6 March 1964) and a production **MiG-25R** recce variant passed state acceptance tests in 1969. The **MiG-25RB** was a dual-role reconnaissance bomber able to drop stores from high altitudes at supersonic speeds. Further models were the **MiG-25RBS**, the **MiG-25RBSh** and **MiG-25RBV**. The first radar recce version was the **MiG-25RBK 'Foxbat-D'**, with SLAR. The Elint **MiG-25RBF** had passive receivers under the nose. MiG-25RBs were exported to Algeria, Bulgaria, India, Iraq, Libya and Syria.

The dedicated **MiG-25RU** recce trainer has no cameras, but like other reconnaissance aircraft has reduced wing span and a constant-sweep leading edge, instead of the fighter's 'cranked' leading edge.

The dedicated defence-suppression **MiG-25BM** model was designed for high-level, long-range, stand-off, anti-radar missions, using four underwing AS-11 'Kilter' missiles. MiG-25BMs often carry the huge underfuselage auxiliary fuel tanks associated with the MiG-25PD.

The MiG-25PD 'Foxbat-E' is the ultimate MiG-25 interceptor and features an IRST and an RP-25 look-down/shoot-down radar.

Mikoyan-Gurevich MiG-25PD 'Foxbat-E'

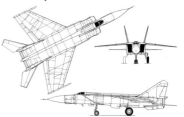

Specification: Mikoyan-Gurevich MiG-25PDS 'Foxbat-E'
Powerplant: two MNPK 'Soyuz' (Tumanskii) R-15BD-300 turbojets each rated at 109.83 kN (24,691 lb st) with afterburning
Dimensions: wing span 14.02 m (45 ft 11.75 in); length 23.82 m (78 ft 1.75 in) or, in aircraft modified with IFR capability, 24.07 m (78 ft 11.67 in); height 6.10 m (20 ft 0.25 in); wing area 61.40 m² (660.93 sq ft)
Weights: normal take-off 34920 kg (76,894 lb); maximum take-off 36720 kg (80,952 lb)
Performance: maximum level speed 'clean' at 13000 m (42,650 ft) Mach 2.8 or 1,619 kt (3000 km/h; 1,864 mph); climb to 20000 m (65,615 ft) in 8 minutes 54 seconds; service ceiling 20700 m (67,915 ft); range with internal fuel 933 nm (1730 km; 1,075 miles) subsonic or 675 nm (1250 km; 776 miles) supersonic
Armament: typical intercept load of two R-40 (AA-6 'Acrid') and four R-60 (AA-8 'Aphid') AAMs, or maximum load of four R-40s (two with SARH and two with IR homing seeker heads)

Mikoyan MiG-27 'Flogger'

Ground attack/tactical strike

The MiG-27 is optimised for low-level attack, introducing a simple fixed intake splitter plate, a short jetpipe, a new PrNK-23K nav/attack system and a Fone laser rangefinder.

Mikoyan-Gurevich MiG-27 'Flogger-H'

The **MiG-27** was developed to remedy the deficiencies of earlier attack 'Floggers'. It introduced fixed, simplified intakes, without variable intake ramps, and an engine with a simpler two-position afterburner nozzle. Fuel economy is improved and weight reduced, at the expense of performance. A new six-barrelled GSh-6-30 30-mm cannon with 260 rounds of ammunition replaced the 23-mm cannon.

Further cannon can be carried underwing, including the SPPU-22 gun pod, whose barrels can be depressed for strafing ground targets in level flight. The full range of guided and unguided bombs and rockets can be carried, including tactical nuclear weapons. The aircraft can also be used in the tactical reconnaissance role. Heavier than any other 'Flogger', the MiG-27 has a redesigned undercarriage with high-pressure tyres and bulged main undercarriage doors.

Development

The **MiG-27** was ordered off the drawing board, and first flew in prototype form in 1972. The 'straight' MiG-27 was soon replaced by the **MiG-27K**, equipped with the PrNK-23K nav/attack system and a Fone laser rangefinder and target tracker. The MiG-27K was capable of highly accurate automatic night/bad weather blind bombing.

There are several sub-variants known as **'Flogger-Js'**, all with a new Klen laser rangefinder and extended LERXes. These improve high-Alpha handling, and mount RWR antennas. The first of the 'Flogger-Js' was the **MiG-27D**, which had an enlarged laser window in the nose. The twin pitot probes serving the nav/attack system are mounted high on the nose. The **MiG-27M 'Flogger-J2'** has an undernose fairing (associated with rearward/downward laser designation) and low-mounted twin pitot probes. Some externally identical aircraft bear the air force designation **MiG-27K**, signifying installation of the Kaira 24 TV/laser weapons designator. India builds the type under licence, manufacturing aircraft to essentially MiG-27D standard. They have the service designation **MiG-27M**, but Mikoyan refers to these as the **MiG-27L**. India requires 165 aircraft to equip six squadrons.

The 'Flogger-D' was the original MiG-27, featuring a plain 'duck-nose' shape. Later variants added nose sensors.

Specification: Mikoyan MiG-27 'Flogger-D'
Powerplant: one MNPK 'Soyuz' (Tumanskii) R-29B-300 turbojet rated at 78.45 kN (17,637 lb st) dry and 112.77 kN (25,353 lb st) with afterburning
Dimensions: wing span 13.97 m (45 ft 9.8 in) spread and 7.78 m (25 ft 6.25 in) swept; length 17.08 m (56 ft 0.25 in) including probe; height 5.00 m (16 ft 5 in); wing area 37.35 m² (402.05 sq ft) spread and 34.16 m² (367.71 sq ft) swept
Weights: empty equipped 11908 kg (26,252 lb); normal take-off 18100 kg (39,903 lb); maximum take-off 20300 kg (44,753 lb)
Performance: maximum level speed 'clean' at 8000 m (26,245 ft) 1,017 kt (1885 km/h; 1,170 mph) or at sea level 728 kt (1350 km/h; 839 mph); combat radius 540 km (291 nm; 335 miles) on lo-lo-lo attack with two Kh-29 ASMs and three drop tanks
Armament: one GSh-6-30 30-mm cannon with 260 rounds, maximum ordnance more than 4000 kg (8,818 lb), including full range of Russian laser-, TV- and EO-guided ASMs, rockets, guided and unguided bombs, UPK-23 23-mm and SPPU-22 23-mm gun pods, recce pods and tactical nuclear weapons

Mikoyan MiG-29 'Fulcrum'

Russia's F-16 equivalent, the MiG-29 has been supplied to several client states and former satellites. India was the most important export customer, receiving 70 'Fulcrums'.

The **MiG-29** was developed to meet a 1972 requirement for a lightweight fighter to replace MiG-21s, MiG-23s and Su-17s. It was to be capable of destroying any enemy fighter in air combat, while having a secondary ground-attack role. Genuine BVR and look-down/shoot-down capability and a capacity for truly autonomous operation were incorporated, together with rough-field capability. Design work began in 1974. The first of 11 prototypes flew on 6 October 1977 and was allocated the provisional reporting name **'Ram-L'**. Operational evaluation began in 1983, and the aircraft entered service soon afterwards.

Armament and avionics
In the fighter role, the MiG-29 **'Fulcrum-A'** carries two BVR AA-10 'Alamo-As' inboard and four short-range AA-8 'Aphid' or AA-11 'Archer' IR-homing missiles outboard, backed by an internal 30-mm cannon. The MiG-29 has two targeting sensors, the N-019 pulse-Doppler radar and, to give a measure of passive capability, an IRST system. This can detect, track and engage a target while leaving the radar in a non-emitting mode. For close-in engagements, a helmet-mounted sight can be used to cue IR-homing missiles onto an off-boresight target. For its secondary ground-attack role, the MiG-29 can carry a variety of bombs (including the 30-kT RN-40 nuclear weapon), rockets and missiles. More than 500 single-seat 'Fulcrums' are estimated to be in service with the VVS, and others have been exported to Bulgaria, Cuba, Czechoslovakia, East Germany, Hungary, India, Iran, Iraq, North Korea, Poland, Romania, Syria and Yugoslavia; 18 have been ordered by Malaysia.

'Fulcrum-C' and trainer variant
NATO allocated the reporting name **'Fulcrum-C'** for aircraft fitted with a bulged spine, which houses both fuel and avionics. None have been exported. The radar-less **MiG-29UB** trainer has a weapons system simulator, allowing the instructor to generate HUD, IRST and radar symbology in the front cockpit. It is in service with most MiG-29 operators.

Supporting the single-seat MiG-29 is the MiG-29UB two-seat conversion trainer variant. Although it lacks NO-193 radar, it does have a sophisticated weapons system simulator.

Mikoyan MiG-29 'Fulcrum-A'

Specification: Mikoyan MiG-29 'Fulcrum-A'
Powerplant: two Klimov/Leningrad RD-33 turbofans each rated at 49.42 kN (11,111 lb st) dry and 81.39 kN (18,298 lb st) with afterburning
Dimensions: wing span 11.36 m (37 ft 3.25 in); length 17.32 m (56 ft 9.85 in); height 4.73 m (15 ft 6.2 in); wing area 38.00 m^2 (409.04 sq ft)
Weights: operating empty 10900 kg (24,030 lb); normal take-off 15240 kg (33,598 lb) as an interceptor; maximum take-off 18500 kg (40,785 lb) in strike configuration
Performance: maximum level speed 'clean' at 11000 m (36,090 ft) 1,319 kt (2445 km/h; 1,519 mph) or at sea level 810 kt (1500 km/h; 932 mph); maximum rate of climb at sea level 19800 m (64,961 ft) per minute; service ceiling 17000 m (55,775 ft); ferry range 1,134 nm (2100 km; 1,305 miles) with three tanks; range 810 nm (1500 km; 932 miles) with internal fuel
Armament: one GSh-301 30-mm cannon, maximum ordnance 3000 kg (6,614 lb), (intercept) two BVR AA-10 and four short-range AA-8 or AA-11 IR-homing missiles, (attack) bombs, rockets and ASMs

Mikoyan MiG-29M/K/S

The MiG-29M demonstrator is seen carrying eight dummy R-77 AAM-AE medium-range missiles. These are believed to have inertial guidance and terminal active radar homing.

The **MiG-29M** designation applies to an advanced multi-role 'Fulcrum' variant which introduces a quadruplex analogue FBW control system, and a two MFD cockpit. The new N-010 radar provides new modes, automatic terrain-following, and increased (25%) detection range. The improved IRST incorporates a TV camera and laser designator for 'smart' weapons. Redesigned LERXes allow 50% more internal fuel and the bulged spine (similar to 'Fulcrum-C') accommodates new chaff/flare dispensers. There are new wingtip ECM/RWR antennas, and extended span ailerons. The tailplane has increased chord and leading edge dogtooth. Powered by uprated RD-33K engines, performance is little changed, although high AoA capability is improved. The MiG-29M flew in mid-1986 and completed acceptance trials, but funding was not obtained.

The similar navalised **MiG-29K** was cancelled when Russia standardised fighter production on Su-27 variants, although the its multi-role capability and smaller size made it the Navy's favourite. It featured an arrester hook, a strengthened undercarriage, folding wings and a retractable inflight-refuelling probe. The increased chord wing has the extended-span ailerons of the MiG-29M, but has extended wingtips, and new, broader-chord double-slotted flaps. Underwing hardpoints carry the same range of AAMs and AGM weapons as the MiG-29M, as well as ASMs.

MiG-29S

Although the second generation MiG-29M and K have received no production funding, Mikoyan are marketing an upgraded derivative of the basic 'Fulcrum-A/C' as the **MiG-29S**. A modified FCS uses computers to improve controllability and high Alpha capability. Software has been improved and processing capacity has been increased. Capability has been enhanced by a new sighting system, and by provision for the active homing AAM-AE. External warload is doubled by restressing the inner underwing pylons. Mikoyan/MAPO will offer an IFR probe, ground mapping radar modes, Western avionics and compatibility with anti-radar, TV- and laser-guided ASMs.

Mikoyan's abortive attempt to provide the Russian navy with a carrierborne fighter was the MiG-29K.

Mikoyan MiG-29K

Specification: Mikoyan MiG-29M 'Fulcrum-?'
Powerplant: two Leningrad/Klimov (Isotov/Sarkisov) RD-33K turbofans rated at 86 kN (19,335 lb st) with afterburning
Dimensions: wing span 11.36 m (37 ft 3.25 in); length approximately 17.37 m (57 ft 0 in) including probe; height 4.73 m (15 ft 6.2 in)
Weights: operating empty 10900 kg (24,030 lb); normal take-off 15240 kg (33,598 lb) as an interceptor; maximum take-off 18500 kg (40,785 lb) in strike configuration
Performance: maximum level speed 'clean' at 11000 m (36,090 ft) 1,319 kt (2445 km/h; 1,519 mph) or at sea level 810 kt (1500 km/h; 932 mph); maximum rate of climb at sea level 19800 m (64,961 ft) per minute; service ceiling 17000 m (55,775 ft); range 1,728 nm (3200 km; 1,988 miles) with external tanks
Armament: one GSh-301 -mm cannon, maximum ordnance 4500 kg (9,921 lb), including AA-8, -10 and -11 AAMs and provision for AAM-AE, compatible with anti-radar, TV- and laser-guided air-to-surface missiles

Mikoyan MiG-31 'Foxhound'

The MiG-31M is a prototype for an improved 'Foxhound' variant with a large radar capable of engaging six targets. This example carries R-77 AAM-AE 'AMRAAMski' missiles.

The **MiG-31 'Foxhound'** was developed to intercept low-level strike aircraft and cruise missiles, complementing the Su-27 in service, using its ultra-long-range capability to fill gaps in Russia's ground-based radar chain. A two-seat derivative of the MiG-25 'Foxbat', with all-new structure, the MiG-31 introduced a new wing planform with small LERXes, Soloviev D-30F-6 turbofans and a new undercarriage. A **Ye-155MP** prototype flew on 16 September 1975 and series production of 280 MiG-31s began in 1979.

The 'Zaslon' radar has a phased-array antenna, increasing range and allowing faster, more accurate beam pointing. It has a detection range of 200 km (125 miles) against a fighter target, and a tracking range of 120 km (75 miles). Ten targets can be tracked simultaneously, and four engaged. Groups of four MiG-31s can operate independently of ground control, covering a 900-km (560-mile) swathe of territory, with the leader automatically controlling his wingmen.

The 'Foxhound' carries a GSh-6-23 six-barrelled 23-mm cannon with two AA-6 'Acrid' or four AA-8 'Aphid' missiles underwing. Primary armament is the R-33 (AA-9 'Amos') long-range SARH missile, four of which are carried under the belly. Later MiG-31s are fitted with a semi-retractable IFR probe, and have four pylons.

Improved variants

The improved **MiG-31M** interceptor variant has been built only in prototype form. Its new radar has a 1.4-m diameter antenna and can simultaneously engage six targets. A fully-retractable IRST is fitted, and MiG-31Ms also have a redesigned rear cockpit, with three CRT MFDs. Other changes include a one-piece canopy and windscreen, a retractable IFR probe, large wingtip ESM pods, and aerodynamic refinements. Redesigned LERXes improve high AoA handling. A bulged spine gives increased fuel capacity.

The MiG-31M carries six long-range AAMs in three side-by-side recesses under the belly, each accommodating tandem pairs of missiles. The similar **MiG-31D** retains the original 1.1-m radar. The designation **MiG-31BS** is applied to MiG-31Ds produced by conversion.

Although based on the MiG-25, the 'Foxhound' has reduced absolute top speed performance but much improved handling.

Mikoyan MiG-31 'Foxhound'

Specification: Mikoyan MiG-31 'Foxhound-A'
Powerplant: two Aviavidgatel D-30F6 turbofans each rated at 151.9 kN (34, 170 lb st) with afterburning
Dimensions: wing span 13.464 m (44 ft 2 in); length 22.69 m (74 ft 5.25 in) including probe; height 6.15 m (20 ft 2.25 in); wing area 61.6 m² (663.0 sq ft)
Weights: empty 21825 kg (48,115 lb); maximum take-off 46200 kg (101,850 lb) with maximum internal fuel and two underwing fuel tanks
Performance: maximum level speed at 17500 m (57,400 ft) 1,620 kt (3000 km/h; 1,865 mph), and at sea level 810 kt (1500 km/h; 9322 mph); service ceiling 20600 m (67,600 ft); combat radius with maximum internal fuel and four R-33 AAMs at Mach 0.85 647 nm (1200 km; 745 miles)
Armament: one GSh-6-23 23-mm cannon with 260 rounds, maximum intercept load of four long-range R-33 (AA-9 'Amos' SARH), two medium-range R-40T (AA-6 'Acrid' IR-homing) or four short-range R-60 (AA-8 'Aphid') AAMs

Mil Mi-8 'Hip'

One of the most common military types in current service, the Mi-8/17 family has been exported to a many countries. India acquired 90 Mi-8s, followed by 50 more-capable Mi-17s.

The **Mi-8** was designed as a turbine-engined Mil Mi-4 derivative, using the same tailboom and rotors. The new Isotov turboshaft was relocated above the fuselage, allowing a simpler transmission and bigger cabin for up to 28 troops. The single-engined prototype (**'Hip-A'**) flew during 1961, followed by the **'Hip-B'**, powered by twin TV2 engines. The **Mil Mi-8P 'Hip-C'** was a passenger/freight transport, while the **Mi-8S** was a passenger airliner with toilet and galley.

Dedicated miltary variants

The **Mil Mi-8T** is a dedicated utility transport with circular cabin windows, and optional outriggers carrying four weapons pylons. The **Mi-8TB 'Hip-E'** is a dedicated assault derivative, with a nose-mounted machine-gun. It has new outriggers with three underslung pylons per side. Above the outer four pylons are launch rails for the AT-2 'Swatter' ATGM. The export **Mil Mi-8TBK 'Hip-F'** had six 'overwing' launch rails for the AT-3 'Sagger' ATGM. To improve performance, the Mi-8 was re-engined with uprated TV3-117MTs to produce the **Mil Mi-17 'Hip-H'**. The new aircraft has PZU intake filters, and the tail rotor is relocated from starboard to port. CIS/Russian air forces use the **Mil Mi-8MT** or **Mi-8TV** designations depending on equipment fit.

Specialised variants

The **'Hip-D'** was a command post/radio relay platform, with a pair of tubular antennas above the rear fuselage, and a V-shaped antenna mast under the tailboom. The **Mil Mi-9 'Hip-G'** is another command post/radio relay 'Hip' variant, with 'hockey stick' antennas under the tailboom. The **Mil Mi-8SMV 'Hip-J'** operates in the ECM jamming role. The **Mil Mi-8PPA 'Hip-K'** is a communications jammer, with box fairings on the fuselage sides, a complex mesh-on-tubular-framework antenna array on the rear fuselage and six side-by-side heat exchangers below the forward fuselage. The **Mi-17P** or **Mi-17PP 'Hip-H (EW)'** has the same heat exchangers and box-like fairings on the fuselage sides but has a solid array in place of the mesh antenna.

The Mi-8 is principally an assault helicopter, but it has given rise to many special mission variants. The Finnish air force received six Mi-8s: four for SAR and two for VIP transport.

Mil Mi-8 'Hip-E'

Specification: Mil Mi-8 'Hip-C'
Powerplant: two Klimov (Isotov) TV2-117A turboshafts each rated at 1257 kW (1,700 shp)
Rotor system: main rotor diameter 21.29 m (69 ft 10.25 in); length overall, rotors turning 25.24 m (82 ft 9.75 in) and fuselage 18.17 m (59 ft 7.35 in) excluding tail rotor; height overall 5.65 m (18 ft 6.5 in); main rotor disc area 356.00 m² (3,832.08 sq ft)
Weights: typical empty 7260 kg (16,007 lb); normal take-off 11100 kg (24,471 lb); maximum take-off 12000 kg (26,455 lb); maximum payload 4000 kg (8,818 lb)
Performance: maximum level speed 'clean' at 1000 m (3,280 ft) 140 kt (260 km/h; 161 mph); maximum cruising speed at optimum altitude 122 kt (225 km/h; 140 mph); service ceiling 4500 m (14,760 ft); hovering ceiling 1900 m (6,235 ft) in ground effect and 800 m (2,625 ft) out of ground effect; ferry range 648 nm (1200 km; 746 miles) with auxiliary fuel; range 251 nm (465 km; 289 miles) with standard fuel

Mil Mi-24 'Hind'

Operational experience with the original Mi-24 'Hind-A' led to redesign of its nose section. The 'Hind-D' (pictured above) introduced a new nose with stepped, armoured cockpits.

Mil Mi-24 'Hind-G'

The **Mil Mi-24** was developed from the Mi-8 'Hip', using the same engines and rotor but intended as a flying APC, carrying a squad of soldiers and providing its own suppressive fire, and relying on speed for protection. A **V-24** prototype was first flown in 1970 and the production **'Hind-A'** entered front-line service during 1973, armed with AT-2 'Swatter' missiles. During production the TV3-117 engine (used by the Mil Mi-17) was introduced, leading to repositioning of the tail rotor to the port side of the tailboom.

'Hind-D' and 'Hind-E'

As the type's transport role declined in importance, anti-tank capability became more important. The cockpit gave inadequate visibility and offered little protection. The solution was an entirely new nose, with separate, stepped, heavily armoured tandem cockpits for the pilot (rear) and gunner (front). Under the nose was a stabilised turret housing a four-barrelled 12.7-mm gun. The new aircraft is designated **Mil Mi-24D 'Hind-D'** (**Mil Mi-25** for export). This was soon replaced by the **Mil Mi-24V 'Hind-E'** (**Mi-35** for export) armed with tube-launched AT-6 'Spiral' missile.

The **Mil Mi-24P** was developed as a result of Afghanistan combat experience. The original 12.7-mm machinegun proved ineffective against some targets, and the obvious answer was a larger-calibre cannon. Accordingly, Mil designed the **Mil Mi-24VP**, a 'Hind-E' with a twin-barrelled GSh-23L in its nose turret, and the **Mil Mi-24P 'Hind-F'** which mounts a GSh-30-2 twin-barrelled 30-mm cannon on the starboard forward fuselage. Export versions of the 'Hind-F' are designated **Mil Mi-25P** and **Mil Mi-35P**.

The **Mil Mi-24RCh 'Hind-G'** undertakes NBC reconnaissance, picking up soil samples to ascertain the extent of contamination, using 'clutching hand' grabs which replace the wingtip missile launch rails. The aircraft is also fitted with a rearward-firing marker flare dispenser mounted on the tailskid. The **Mil Mi-24K 'Hind-G2'** is based on the Mi-24RCh, but has a large, bulky camera housing under the nose, offset to starboard, and an even larger camera (starboard oblique) in the cabin.

Informally known in the West as the 'Devil's Chariot', the Mi-24 is a fearsome combined assault/attack helicopter.

Specification: Mil Mi-24D 'Hind-D'
Powerplant: two Klimov (Isotov) TV3-117 Series III turboshafts each rated at 1640 kW (2,200 shp)
Dimensions: main rotor diameter 17.30 m (56 ft 9 in); wing span 6.536 m (21 ft 5.5 in); length overall, rotors turning 19.79 m (64 ft 11 in) and fuselage 17.51 m (57 ft 5.5 in) excluding rotors and gun; height overall 6.50 m (21 ft 4 in) with rotors turning; main rotor disc area 235.00 m² (2,529 sq ft)
Weights: empty 8400 kg (18,519 lb); normal take-off 11000 kg (24,250 lb); maximum take-off 12500 kg (27,557 lb)
Performance: maximum level speed 'clean' at optimum altitude 168 kt (310 km/h; 192 mph); maximum rate of climb at sea level 750 m (2,461 ft) per minute; service ceiling 4500 m (14,765 ft); combat radius 86 nm (160 km; 99 miles) with maximum military load
Armament: one four-barrelled JakB 12.7-mm gun, maximum ordnance 2400 kg (5,291 lb), including up to four AT-2 'Swatter' ATGMs and four UV-32-57 57-mm rocket pods

Mil Mi-28 'Havoc'

One of the Mi-28 prototypes displays the huge 30-mm cannon mounted under the nose. AT-6 'Spiral' anti-tank guided missiles are carried on the stub pylons.

The first of three **Mil Mi-28** attack helicopter prototypes made its maiden flight on 10 November 1982. The Mi-28 is of conventional configuration, with an undernose cannon and stepped armoured cockpits accommodating pilot (rear) and gunner (forward). A conventional three-bladed tail rotor was abandoned and replaced on the second and third prototypes by a 'scissor'-type tail rotor, with two independent two-bladed rotors on the same shaft, set at approximately 35° to each other and forming a narrow X.

Cannon and missile armament

The Mil Mi-28 is armed with a single-barrelled 2A42 30-mm cannon under the nose, with twin 150-round ammunition boxes co-mounted to traverse, elevate and depress with the gun itself, reducing the likelihood of jamming. The stub wings have four pylons, each able to carry 480 kg (1,058 lb), typically consisting of four tube-launched AT-6 'Spiral' missiles or a variety of rocket pods. The wingtip houses a chaff/flare dispenser.

Armour and crash protection

The cockpit is covered by flat, non-glint panels of armoured glass, and is protected by titanium and ceramic armour. Vital components are heavily protected and duplicated, and shielded by less important items. In the event of a catastrophic hit the crew are protected by energy absorbing seats, which can withstand a 12 m/sec crash landing. An emergency escape system is installed which blows off the doors and inflates air bladders on the fuselage sides. The crew roll over these before pulling their parachute ripcords.

A hatch in the port side, aft of the wing, gives access to the avionics compartment and to an area large enough to accommodate two or three people. This is intended to allow a Mi-28 to pick up the crew of another downed helicopter. The production **Mil Mi-28N** will feature FLIR and LLLTV in the nose, on each side of the laser rangefinder/designator turret, and will have an NVG-compatible cockpit. A transport derivative of the Mi-28, apparently designated **Mi-40**, is under development.

The Mi-28 is highly manoeuvrable, and can bring its formidable weaponry to bear rapidly on to a target.

Mil Mi-28 'Havoc'

Specification: Mil Mi-28 'Havoc-A'
Powerplant: two Klimov (Isotov) TV3-117 turboshafts each rated at 1640 kW (2,200 shp)
Dimensions: main rotor diameter 17.20 m (56 ft 5 in); wing span 4.87 m (16 ft 0 in); length overall, rotors turning 19.15 m (62 ft 10 in) and fuselage 16.85 m (55 ft 3.5 in); main rotor disc area 232.35 m² (2,501.10 sq ft)
Weights: empty 7000 kg (15,432 lb); maximum take-off 10400 kg (22,928 lb)
Performance: maximum level speed 'clean' at optimum altitude 162 kt (300 km/h; 186 mph); maximum cruising speed at optimum altitude 146 kt (270 km/h; 168 mph); service ceiling 5800 m (19,025 ft); hovering ceiling 3600 m (11,810 ft) out of ground effect; range 253 nm (470 km; 292 miles) with standard fuel; endurance 2 hours
Armament: one single-barrelled 2A42 30-mm cannon with two 150-round drums, maximum ordnance about 1920 kg (4,233 lb), including up to 16 AT-6 'Spiral' ATGMs and a variety of rocket pods

Mitsubishi T-2/F-1

Advanced trainer/close-support fighter

The Mitsubishi F-1 close-support fighter has anti-ship attack as its primary role, armed with ASM-1 missiles. Three JASDF squadrons are equipped with the type.

Mitsubishi F-1 and T-2 (side view)

Japan's first excursion into supersonic military aircraft design was of a two-seat combat trainer which would, it was argued, double as an aircraft in which JASDF pilots could be trained for the Lockheed F-104J and McDonnell Douglas F-4EJ combat aircraft, and which would provide design experience for a subsequent indigenous fighter.

First flown on 20 July 1971 as the **XT-2**, the **Mitsubishi T-2** tandem two-seat trainer features shoulder-mounted wings with anhedral and fixed-geometry lateral air intakes for two licence-built Rolls-Royce/Turboméca Adour Mk 801A afterburning turbofans, mounted side by side in the lower rear fuselage. The T-2 is equipped with nose-mounted J/AWG-11 search and ranging radar and licence-built American avionics. Total production orders amounted to 90 aircraft, of which 28 are T-2 advanced trainers (entering service in 1976) and 62 are **T-2A** combat trainers.

One T-2 has been extensively modified as the FBW control-configured vehicle **T-2CCV** with vertical and horizontal canard surfaces, plus test equipment in the rear cockpit

F-1 close-support fighter

Mitsubishi then developed a close-support fighter variant, designated **F-1**, and tasked primarily with anti-shipping missions. Two T-2 prototypes were modified, the first such conversion making its maiden flight on 3 June 1975, followed by the first production F-1 in June 1977. The F-1 is dimensionally identical to the T-2, but the space behind the pilot's (front) cockpit of the trainer has been replaced by a 'solid' fairing containing an avionics compartment with a J/ASQ-1 fire-control system and bombing computer, an INS and a radar warning and homing sub-system. Mitsubishi's J/AWG-12 radar gives compatibility with the F1's principal anti-shipping weapon – the indigenously-developed Mitsubishi ASM-1 missile with active radar terminal seeker. The F1 entered service in April 1978 and the last of 77 aircraft was delivered in March 1987. As a result of delays to the FS-X fighter programme, the F-1s are expected to remain in service until at least 1999/2000 and are undergoing a SLEP which extends airframe life from 3,500 to 4,000 hours.

Still fulfilling a vital advanced training role in the JASDF, the T-2 has seen service as an aggressor in air combat training.

Specification: Mitsubishi F-1
Powerplant: two Ishikawajima-Harima TF40-IHI-801 (Rolls-Royce/Turboméca Adour Mk 801A) turbofans each rated at 22.75 kN (5,115 lb st) dry and 32.49 kN (7,305 lb st) with afterburning
Dimensions: wing span 7.88 m (25 ft 10.25 in); length 17.86 m (58 ft 7 in) including probe; height 4.39 m (14 ft 5 in); wing area 21.17 m² (227.88 sq ft)
Weights: operating empty 6358 kg (14,017 lb); normal take-off 12800 kg (28,219 lb); maximum take-off 13700 kg (30,203 lb)
Performance: maximum level speed 'clean' at 10975 m (36,000 ft) 917 kt (1,056 mph; 1700 km/h); maximum rate of climb at sea level 10670 m (35,000 ft) per minute; service ceiling 15240 m (50,000 ft); combat radius 300 nm (555 km; 345 miles) on a hi-lo-hi attack mission with two ASMs and two tanks
Armament: one JM61 Vulcan 20-mm rotary cannon; maximum ordnance 2721 kg (6,000 lb) including two ASM-1 missiles, four AIM-9L AAMs; plus 227-kg (500-lb) or 340-kg (750-lb) bombs and 19 x 70-mm JLAU-3A, 7 x 70-mm RL-7 and 4 x 125-mm RL-4 rocket pods, or three drop tanks

Nanchang A/Q-5 'Fantan'

A Q-5III displays the wing planform which it inherited from the MiG-19. The Q-5's relative lack of capability in performance and equipment is offset by its low price.

Nanchang Q-5III (A-5C)

Development of the **Nanchang Q-5 'Fantan'** began in 1958 to meet a PLA requirement for a dedicated attack aircraft. Although based on the MiG-19, Nanchang's design retained only the rear fuselage and main undercarriage, introducing a new, stretched, area-ruled fuselage with an internal weapons bay, new conical-section nose, wings of greater area and less sweep, larger tailplanes and lateral air intakes. A prototype made its delayed maiden flight on 4 June 1965, but extensive modifications proved necessary to solve problems with the hydraulics, brakes, fuel and weapons systems. Two new prototypes flew in October 1969 and the type was ordered into production.

Q-5 variants

Little is known of a dedicated Q-5 nuclear weapons-carrier version, which carries a single 5- to 20-kT weapon. The **Q-5I** is an extended-range variant with a new ejection seat, two additional hardpoints and a new Wopen WP6 engine. Some Q-5Is were modified to serve as missile-carriers with the PLA navy, and some of these may have Doppler nose radar. C-801 AShMs and torpedoes could also be carried. The **Q-5IA**, certificated for production in 1985, was fitted with an additional underwing hardpoint and introduced a new gun/bomb sighting system and new defensive avionics. The **Q-5II** received an RWR but was otherwise similar. The **A-5C** (**Q-5III**) was an export Q-5IA for Pakistan with substantially improved avionics and compatibility with AIM-9 AAMs.

Exports and upgrades

Production of the Q-5 continues, and over 1,000 are believed to have been delivered. It has been exported to Bangladesh, North Korea and Pakistan. Several programmes have been launched to upgrade Q-5s with Western avionics and/or equipment. Although yielding flying protoypes, further development was terminated after 1989. The **A-5K Kong Yun** (Cloud) was equipped with a laser rangefinder. The **A-5M** programme added a ranging radar, an INS, a HUD and new IFF and RWR equipment. An extra wing hardpoint was added, along with compatibility with the PL-5 AAM.

Pakistan was a major customer for the export A-5, equipping its aircraft with some Western avionics and weapons.

Specification: Nanchang Q-5 IA 'Fantan'
Powerplant: two Liming (LM) (previously Shenyang) Wopen-6A turbojets each rated at 29.42 kN (6,614 lb st) dry and 36.78 kN (8,267 lb st) with afterburning
Dimensions: wing span 9.68 m (31 ft 9 in); length 15.65 m (51 ft 4.25 in) including probe; height 4.333 m (14 ft 2.75 in); wing area 27.95 m² (300.86 sq ft)
Weights: empty 6375 kg (14,054 lb); normal take-off 9486 kg (20,913 lb); maximum take-off 11830 kg (26,080 lb)
Performance: maximum level speed at 11000 m (36,000 ft) Mach 1.12/643 kt (1190 km/h; 740 mph); maximum speed at sea level 653 kt (1210 km/h; 752 mph); maximum rate of climb at 5000 m (16,400 ft) 4980-6180 m (16,340-20,275 ft) per minute; service ceiling 15850 m (52,000 ft); combat radius with maximum external stores, afterburners off lo-lo-lo 216 nm (400 km; 248 miles)
Armament: two internal 23-mm cannon with 100 rpg; maximum ordnance 2000 kg (4,409 lb) including 250-kg (551-lb), 500-kg (1,102-lb) bombs, practice bombs and rocket pods, PL-5 AAMs

Northrop B-2 Spirit

Stealthy strategic bomber

The B-2 employs split surfaces on the outer wing panels to act as 'drag rudders'. These provide aerodynamic braking and yaw control.

The **Northrop B-2 Spirit** flying wing was developed in great secrecy as a stealthy, or radar-evading, strategic bomber for the Cold War mission of attacking Soviet strategic targets with stand-off nuclear weapons. The B-2 began as a 'black' programme, known in its infancy as **Project Senior C.J.** and later as the **ATB** (Advanced Technology Bomber). At the height of the Cold War, the USAF expected to procure no fewer than 132 B-2s.

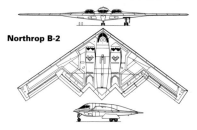

Northrop B-2

Stealthy shape
The B-2's four F118 turbofans are non-afterburning varioants of the F110 turbofan and have intakes and exhausts located atop the aircraft to shield them from detection from below. The crew/payload section of the aircraft starts aft of the apex of the wing, ends at the wing trailing edge and is smoothly blended on the upper surfaces of the wing. The crew compartment provides side-by-side seating for two crew members, both pilots, seated in zero-zero, vertically-ejecting ACES II seats.

USAF officers claim that the B-2's radar-eluding qualities mean that it would not have to dash in and out of enemy airspace quickly like other bombers. Instead, a B-2 could get updates from reconnaissance satellites on positions of mobile targets and avoid being shot down while the two-man crew looked for the targets. To verify targets at the last moment, the B-2 will briefly turn on a special radar that spotlights only a small area, and then attack with nuclear bombs and SRAM II missiles.

Service history
The first flight of the B-2 took place on 17 July 1989. Test flying to evaluate low observables technology began on 30 October 1990. Flight testing of the B-2 was scheduled to continue until 1995, with critical radar cross-section evaluations in 1993. The USAF received is first operational B-2 in December 1993, and expects to achieve IOC with the 509th Bomb Wing at Whiteman AFB, MO, by late 1995. Total procurement, once reduced from 132 to 75, has been curtailed to just 20 front-line aircraft.

The B-2 houses its weapon load in two bays in the central 'fuselage' section immediately behind the crew area.

Specification: Northrop B-2A
Powerplant: four General Electric F118-GE-110 non-afterburning turbofans each rated at 84.52 kN (19,000 lb st)
Dimensions: span 52.43 m (172 ft 0 in); length 21.03 m (69 ft 0 in); height 5.18 m (17 ft 0 in); wing area more than 464.50 m² (5,000.0 sq ft)
Weights: empty between 45360 and 49900 kg (100,000 and 110,000 lb); normal take-off 168433 kg (371,330 lb); maximum take-off 181437 kg (400,000 lb)
Performance: maximum level speed at high altitude about 416 kt (764 km/h; 475 mph); service ceiling 15240 m (50,000 ft); range with a 10886-kg (24,000-lb) warload comprising eight SRAMs and eight B61 bombs 6,600 nm (12231 km; 7,600 miles) with internal fuel on a hi-hi-hi mission or 4,500 nm (8339 km; 5,182 miles) with internal fuel on a hi-lo-hi mission
Armament: maximum ordnance 50,000 lb (22680 kg), including up to 16 AGM-169 SRAM II or AGM-129 ACMs, or up to 80 Mk 82, or 16 Mk 84 bombs, or 36 M117 fire bombs, 36 CBU-87 cluster bombs and 80 Mk 36 or Mk 62 sea mines

Northrop F-5A Freedom Fighter

Canada is one of the last major operators of the early F-5 generation, flying its upgraded CF-116A and CF-116D aircraft on fighter lead-in training and aggressor duties.

In 1954 the US government initiated a study to determine requirements for a simple lightweight fighter to be supplied via the Military Assistance Program. Northrop's private-venture **N-156C** design made its first flight on 30 July 1959, and was selected in 1962 by the USAF as the required 'FX' fighter. It was designated **F-5**, and an **F-5A** prototype was first flown in May 1963. The corresponding two-seat **F-5B** fighter/trainer entered service in April 1964, four months ahead of the F-5A. Northrop also developed the reconnaissance **RF-5A**, equipped with four nose-mounted cameras.

First generation F-5s were exported to Brazil, Greece, Jordan, Morocco, Philippines, Saudi Arabia, South Korea, Spain, Thailand, Turkey, Venezuela and Yemen. Some remain active, but many will soon be stored or scrapped.

Canadian production

In 1965 the F-5 was selected for licence-production by Canadair for the Canadian Armed Forces (CAF). It built both single-seat **Canadair CF-5A**s and **CF-5D** trainers. Several major improvements were incorporated, including uprated engines and an IFR probe. The first CF-5A flew in May 1968, and the type entered service with the CAF later that year, designated **CF-116**. Soon after manufacture started, the Royal Netherlands air force (KLu) ordered 105 **NF-5A**s with leading-edge manoeuvre flaps and Doppler radar. The NF-5 entered service in 1969 but has now been retired.

In 1991 Bristol Aerospace was awarded a contract to upgrade the CF-5 for continued service as a lead-in weapons trainer for the CF-18 Hornet force. Modification involves airframe refurbishment and strengthening (wings, fins, control surfaces and replacement of undercarriage), while an avionics upgrade adds a HUD/weapons aiming system, a laser INS, a digital databus, HOTAS controls, radar altimeter and new radio. Following the maiden flight of a refurbished two-seat CF-5D on 14 June 1991, a total of 11 CF-5As and all 33 surviving CF-5Ds are to be modified.

Surplus CAF and KLu CF-5s and NF-5s have been purchased by Greece, Turkey and Venezuela to augment their existing fleets.

Norway's F-5As received corrosion-proofing and RWRs in the mid-1980s, and augment F-16s in the fighter-bomber role.

Northrop F-5A

Specification: Northrop F-5A Freedom Fighter
Powerplant: two General Electric J85-GE-13 turbojets each rated at 12.10 kN (2,720 lb st) dry and 18.15 kN (4,080 lb st) with afterburning
Dimensions: wing span 7.70 m (25 ft 3 in) without tip tanks and 7.87 m (25 ft 10 in) with tip tanks; length 14.38 m (47 ft 2 in); height 4.01 m (13 ft 2 in); wing area 15.79 m² (170.00 sq ft)
Weights: empty equipped 3667 kg (8,085 lb); maximum take-off 9379 kg (20,677 lb)
Performance: maximum level speed 'clean' at 10975 m (36,000 ft) 802 kt (1487 km/h; 924 mph); cruising speed at 10975 m (36,000 ft) 556 kt (1030 km/h; 640 mph); maximum rate of climb at sea level 8748 m (28,700 ft) per minute; service ceiling 15390 m (50,500 ft); combat radius 485 nm (989 km; 558 miles) on a hi-lo-hi attack mission with two 240-kg (530-lb) bombs and maximum fuel
Armament: two M39 20-mm cannon and two AIM-9 AAMs on wingtip launchers; maximum ordnance of 1996 kg (4,400 lb), including Mk 80 series bombs, rocket-launcher pods, and up to three 568-litre (125-Imp gal) drop tanks

Northrop F-5E Tiger II

Chile's F-5Es have been upgraded by IAI with new cockpits with advanced displays, more capable radars, and capability for Python 3 missiles and laser-guided bombs.

The **Northrop F-5E/F Tiger II** lightweight fighter became the principal US export fighter in the 1970s. It was derived from the earlier F-5A/B, incorporating uprated J85 engines, an integrated fire control system, additional fuel and a larger, modified wing with LERXes and manoeuvring flaps for enhanced manoeuvrability. The **F-5E** is the single-seat variant and was first flown on 11 August 1972. The **F-5F** tandem two-seat trainer features a lengthened fuselage and retains full combat capability. Initial deliveries were made to the USAF in 1973, to prepare the aircraft for foreign users. F-5E/Fs later served for aggressor training with the USAF until the late 1980s. The USN also acquired a small number for its 'adversary' DACT mission.

Some 1,300 F-5E/Fs were later supplied to 20 air forces. Current operators are Bahrain, Brazil, Chile, Honduras, Indonesia, Iran, Jordan, Kenya, Malaysia, Mexico, Morocco, Saudi Arabia, Singapore, South Korea, Sudan, Switzerland, Taiwan, Thailand, Tunisia, USN and Yemen. It was also built under licence in South Korea, Switzerland and Taiwan.

Upgrades

The considerably upgraded GE F404-powered **F-20 Tigershark** (**F-5G**) was rendered superfluous by the availability of F-16As for export. Recently, however, F-5 upgrade programmes have become available. Chilean F-5Es, for example, are being upgraded with EL/M-2032B multi-mode radar (originally developed for the Lavi), improved RWRs, a new HUD, HOTAS controls, colour MFDs and a new INS.

Recce variant

Production was approved in 1978 for a specialised tactical reconnaissance version designated **RF-5E Tigereye**, retaining full combat capability. A modified lengthened nose houses a single camera. This can be augmented by two interchangeable pallets, containing combinations of panoramic cameras and an IR linescanner. The first production aircraft flew in 1982, and the RF-5E has been exported to Malaysia and Saudi Arabia. Six Singaporean F-5Es have been converted to this recce configuration.

The RF-5E Tigereye is a dedicated reconnaissance variant with cameras in the nose in place of radar.

Northrop F-5E Tiger II

Specification: Northrop F-5E Plus Tiger III
Powerplant: two General Electric J85-GE-21B turbojets each rated at 15.5 kN (3,500 lb st) dry and 22.24 kN (5,000 lb) with afterburning
Dimensions: wing span 8.53 m (28 ft 0 in) with tip-mounted AAMs; length 14.45 m (47 ft 4.75 in) including probe; height 4.08 m (13 ft 4.5 in); wing area 17.28 m^2 (186.00 sq ft)
Weights: empty 4349 kg (9,558 lb); maximum take-off 11187 kg (24,664 lb)
Performance: maximum level speed 'clean' at 10975 m (36,000 ft) 917 kt (1700 km/h; 1,056 mph); cruising speed at 10975 m (36,000 ft) 562 kt (647 mph; 1041 km/h); maximum rate of climb at sea level 10455 m (34,300 ft) per minute; service ceiling 15590 m (51,800 ft); combat radius 760 nm (1405 km; 875 miles) with two AIM-9 AAMs
Armament: two M39A2 20-mm revolver cannon with 280 rpg; maximum ordnance 3175 kg (7,000 lb), including rockets, ASMs, AAMs (AIM-9P, Python 3 and Shafrir 2), plus provision for LGBs

Panavia Tornado IDS/ECR

Wearing desert pink camouflage, this RAF Tornado GR.Mk 1 carries two Alarm anti-radiation missiles. These were used to devastating effect to suppress Iraqi radars during Desert Storm.

The VG **Panavia Tornado**, initially known as the Multi-role Combat Aircraft (**MRCA**), was designed to fulfil an Anglo-German-Italian requirement. It was to undertake interdiction, counter-air, close air support, reconnaissance, maritime attack and point interception roles. The prototype flew on 14 August 1974. The **Tornado IDS** (interdictor/ strike) variant penetrates enemy defences at high subsonic speeds, masked from radar detection by automatic, all-weather terrain following, and is protected by a broad spectrum of active and passive self-defence aids.

Service

RAF orders have totalled 228 **GR.Mk 1**s including 14 new reconnaissance aircraft. Two **GR.Mk 1B** squadrons will receive a maritime attack tasking with BAe Sea Eagle antiship missiles and 'buddy' refuelling pods, to replace the RAF's Buccaneers. A mid-life update later in the 1990s will produce the **GR.Mk 4** with a new HUD, updated weapon control system, colour HDD and an improved EW suite. In Germany, the Luftwaffe received 212 IDS aircraft and the Marineflieger received 112. Italy received 100 Tornado IDS, and 48 aircraft were delivered to the Royal Saudi Arabian Air Force. The follow-on 1993 Al-Yamamah II contract covered an additional purchase of 48 IDS-configured aircraft.

Reconnaissance versions

Although the German navy and Italian air force reconnaissance requirements were initially met with a simple multisensor pod, the Luftwaffe and RAF later opted for a more sophisticated variant. The German **Tornado ECR** (Electronic Combat and Reconnaissance) variant incorporates defence suppression capability with an emitter location system, and two AGM-88 HARM missiles under the fuselage. An IR linescan is mounted in a blister under the forward fuselage with a FLIR immediately ahead. Italy elected to produce 16 of its own ECRs by converting existing aircraft. The RAF's 30 **GR.Mk 1A**s have no specific defence-suppression capability, but may revert to a secondary attack role. They are equipped with an IR linescan.

The only export order for the Tornado came from Saudi Arabia, which bought a mix of GR.Mk 1s and F.Mk 3s.

Tornado GR.Mk 1 and GR.Mk 1A (side view)

Specification: Panavia Tornado GR.Mk 1
Powerplant: two Turbo-Union RB.199 Mk 103 turbofans each rated at 38.48 kN (8,650 lb st) dry and 71.50 kN (16,075 lb st) with afterburning
Dimensions: wing span 13.91 m (45 ft 7.5 in) minimum sweep and 8.60 m (28 ft 2.5 in) maximum sweep; length 16.72 m (54 ft 10.25 in); height 5.95 m (19 ft 6.25 in); wing area 26.60 m² (286.33 sq ft)
Weights: basic empty about 13890 kg (30,620 lb); operating empty 14091 kg (31,065 lb); normal take-off 20411 kg (45,000 lb); maximum take-off about 27951 kg (61,620 lb)
Performance: limiting IAS 800 kt (1482 km/h; 921 mph); service ceiling more than 15240 m (50,000 ft); combat radius 750 nm (1390 km; 863 miles) on a typical hi-lo-hi attack mission
Armament: two 27-mm IWKA-Mauser cannon with 180 rpg; maximum ordnance over 9000 kg (19,841 lb) including the WE177B nuclear bomb, JP233 airfield-denial weapon, 1,000-lb free-fall, retarded and Paveway II laser-guided bombs

Panavia Tornado ADV

During Desert Storm, Saudi Tornado ADVs were integrated into Saudi Arabia's multi-national air defence network but, like their RAF F.Mk 3 equivalents, achieved no aerial successes.

Developed from the Tornado interdictor for wholly British requirements, the **Tornado ADV (Air Defence Variant)** is optimised for long-range interception. Its primary missions are the protection of NATO's northern and western approaches, and long-range air defence of UK maritime forces. Design of the ADV was based around the need to carry four semi-recessed Sky Flash radar-homing AAMs, resulting in a lengthened airframe which also provides increased internal fuel capacity.

The AI.Mk 24 Foxhunter intercept radar is capable of detecting targets at more than 185 km (115 miles) and tracking 20 while continuing to scan. The subject of serious development problems, it was almost cancelled before being brought to a minimally acceptable standard in 1989.

Three ADVs emerged as non-operational prototypes (first flying on 27 October 1979) and 18 as interim standard **Tornado F.Mk 2**s with RB.199 Mk 103 engines. These early aircraft initially flew with lead ballast in place of delayed radars and served only for operational training duties until January 1988. Plans for their conversion to near Mk 3 standard as **F.Mk 2A**s have not been implemented.

Definitive Tornado F.Mk 3 model

The **Tornado F.Mk 3** is the current definitive production version and first flew in November 1985. BAe produced 144 aircraft, including 38 dual-control F.Mk 3s. A cancelled Omani order for eight was later transferred to the RAF. The F.Mk 3 introduces Mk 104 engines, a second INS, and provision for four AIM-9s. The Stage 1 upgrade is introduced in 1989 and retrofitted throughout the fleet. Provisions include a new 'combat stick' with HOTAS controls, radar and Hermes RHAWS improvements, a five per cent combat boost switch for engines, and VICON 78 flare dispensers below the rear fuselage (from 1991). The planned Stage 2 upgrade, in the mid-1990s, will be concerned mainly with further radar improvements.

Seven front-line RAF units currently operate the Tornado F.Mk 3. Saudi Arabia is the sole export customer, receiving 24 from 1989. Both RAF and RSAF Tornado interceptors participated in the Gulf War without seeing aerial combat.

A Tornado F.Mk 3 displays its heavy missile armament.

Panavia Tornado F.Mk 3

Specification: Panavia Tornado F.Mk 3
Powerplant: two Turbo-Union RB.199-34R Mk 104 turbofans each rated at 40.48 kN (9,100 lb st) dry and 73.48 kN (16,520 lb st) with afterburning
Dimensions: wing span 13.91 m (45 ft 7.5 in) spread and 8.60 m (28 ft 2.5 in) swept; length 18.68 m (61 ft 3.5 in); height 5.95 m (19 ft 6.25 in); wing area 26.60 m² (286.33 sq ft)
Weights: operating empty 14502 kg (31,970 lb); maximum take-off 27986 kg (61,700 lb)
Performance: maximum level speed 'clean' at 36,000 ft (10975 m) 1,262 kt (1,453 mph; 2338 km/h); operational ceiling about 70,000 ft (21335 m); combat radius more than 300 nm (345 miles; 556 km) supersonic or more than 1,000 nm (1,151 miles; 1,852 km) subsonic; endurance 2 hours on a CAP at between 300- and 400-nm (345- and 460-mile; 555- and 740-km) radius
Armament: one internal 27-mm Mauser cannon, plus four semi-recessed BAe Sky Flash SARH medium-range missiles and four AIM-9L/M IR-homing short-range missiles

Pilatus PC-7 and PC-9

The PC-9 is a sophisticated PC-7 derivative, featuring a more advanced cockpit design and an uprated PT6 engine. Saudi Arabia was the first customer, ordering 30 PC-9s in 1985.

Pilatus PC-9

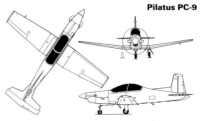

In 1973, Pilatus modified a Swiss air force piston-engined P-3 trainer with a PT6A-25 turboprop engine. It first flew in 1975 but then underwent major structural changes to take full advantage of the extra power. Pilatus introduced a new wing with integral tanks, a new undercarriage and a bubble canopy for the non-pressurised cockpit. For weapons training, six underwing hardpoints can carry external stores up to 1040 kg (2,293 lb). The first production **PC-7 Turbo Trainer** made its maiden flight on 18 August 1978. First deliveries were to Burma in 1979. With the Beech T-34C then its sole production competitor, the PC-7 achieved growing export success, supplemented (after a two-year evaluation) in June 1981 by a Swiss order for 40.

Nearly 440 PC-7s had been sold by 1993 to 18 countries, identified military customers are Abu Dhabi, Angola, Austria, Bolivia, Burma, Chad, Chile, France, Guatemala, Iran, Iraq, Malaysia, Mexico, the Netherlands, Switzerland and Surinam. PC-7s are believed to have been used operationally by both sides in the Iran/Iraq war. In 1985, Pilatus offered an optional installation of twin 0-m/60-kt Martin-Baker CH Mk 15A lightweight ejection seats.

PC-9

The prototype **Pilatus PC-9** made its first flight on 7 May 1984 and was powered by an 857-kW (1,150 shp) P&WC PT6A-62 turboprop. Although derived from the PC-7, there is only 10 per cent structural commonality between the two. The PC-9 is dimensionally similar, but is distinguished by its larger canopy, 'stepped' tandem cockpits with ejection seats, ventral airbrake and four-bladed propeller.

The PC-9 has been exported to Myanmar, Saudi Arabia and Thailand. Most of Australia's 67 **PC-9A**s are assembled under licence and have a Bendix EFIS. A **PC-9B** target-towing version , is operated on behalf of the Luftwaffe.

Pilatus is offering the **PC-9 Mk 2** in conjunction with Beechcraft as a contender for the USAF/USN JPATS competition. It features almost 70 per cent redesign, introducing a strengthened fuselage and a pressurised cockpit with new digital avionics. The prototype first flew in 1992.

Ten PC-7s were delivered to the Royal Netherlands air force in 1989. They are operated on elementary training duties.

Specification: Pilatus PC-7 Turbo-Trainer
Powerplant: one 485-kW (650-shp) Pratt & Whitney Canada PT6A-25A turboprop flat-rated at 410 kW (550 shp)
Dimensions: wing span 10.40 m (34 ft 1 in); length 9.78 m (32 ft 1 in); height 3.21 m (10 ft 6 in); wing area 16.60 m² (178.69 sq ft)
Weights: basic empty 1330 kg (2,932 lb); normal take-off 1900 kg (4,188 lb) for aerobatics; maximum take-off 2700 kg (5,952 lb)
Performance: never exceed speed 270 kt (500 km/h; 311 mph); maximum cruising speed at 6095 m (20,000 ft) 222 kt (412 km/h; 256 mph); maximum rate of climb at sea level 655 m (2,150 ft) per minute; climb to 5000 m (16,400 ft) in 9 minutes; service ceiling 10060 m (33,000 ft); take-off distance to 15 m (50 ft) 1180 m (3,870 ft) at MTOW; landing distance from 15 m (50 ft) 800 m (2,625 ft) at maximum landing weight; range 647 nm (1200 km; 746 miles); endurance 4 hours 22 minutes
Armament: maximum ordnance 1040 kg (2,293 lb) including gun pods, rockets, light bombs

Rockwell B-1B Lancer

The B-1 is optimised for ultra-low-level penetration at high subsonic speeds. The small vanes provide yaw and pitch damping to smooth the otherwise turbulent low-level ride.

Suffering one of the more protracted development periods of any recent military aircraft, today's **Rockwell B-1B Lancer** long-range multi-role strategic bomber is derived from the preceding **B-1A** design. The B-1A programme was cancelled in 1977, and then resurrected in September 1981 with an order for 100 B-1Bs. The design features a blended low-wing/body configuration with VG outer wing panels and advanced high-lift devices, four GE F101 turbofans (mounted in pairs below the wing) with fixed intake geometry, a strengthened landing gear, three internal weapons bays, a moveable bulkhead in the forward weapons bay to allow for the carriage of a diverse range of different sized weapons, optional weapons bay fuel tanks for increased range and external under fuselage stores stations for additional fuel or weapons. The low-altitude, high-speed penetration role against sophisticated air defence systems was to be carried out using electronic jamming equipment, IR countermeasures, radar location and warning systems and application of 'low observable' technology.

The offensive avionics system is centred around the AN/APQ-164 multi-mode radar, which includes a low-observable phased-array antenna for low-altitude terrain following and accurate navigation. The much-troubled AN/ALQ-161 system forms the core of the Lancer's continuously upgradeable defensive capability.

Service and conventional role

The first production B-1B flew on 18 October 1984. Deliveries began on 27 July 1985 with SAC achieving IOC exactly a year later, thereafter rapidly building up a force of four bomb wings. Since then, the career of the B-1B has been coloured by much controversy and interrupted by frequent lengthy grounding orders, and several highly-publicised losses. Thus far, the Lancer has been primarily concerned with strategic applications and is compatible with a variety of nuclear devices, which it can deliver over an unrefuelled range of approximately 12000 km (7,455 miles). An expansion of conventional capability is imminent.

The B-1 acquires a sleek, purposeful look with its wings fully swept. Each wing has high-lift devices to enable the B-1 to become airborne from alert pads in under three minutes.

Rockwell B-1B Lancer

Specification: Rockwell B-1B Lancer
Powerplant: four General Electric F101-GE-102 turbofans each rated at 64.94 kN (14,600 lb st) dry and 136.92 kN (30,780 lb st) with afterburning
Dimensions: wing span 41.67 m (136 ft 8.5 in) (15°) and 23.84 m (78 ft 2.5 in) (67° 30'); length 44.81 m (147 ft 0 in); height 10.36 m (34 ft 10 in); wing area approximately 181.16 m² (1,950.00 sq ft)
Weights: empty equipped 87091 kg (192,000 lb); maximum take-off 216365 kg (477,000 lb)
Performance: maximum level speed 'clean' at high altitude about Mach 1.25 or 715 kt (1324 km/h; 823 mph); penetration speed at 61 m (200 ft) more than 521 kt (965 km/h; 600 mph); range 6,475 nm (12000 km; 7,455 miles) with standard fuel
Armament: inernal maximum payload of 34020 kg (75,000 lb), including AGM-69A SRAM-As; B-28, B-61 or B-83 free-fall nuclear bombs; AGM-86B ALCMs (and possibly AGM-129A ACMs); 84 227-kg (500-lb) Mk 82 bombs or 227-kg (500-lb) Mk 36 mines internally, AGM-86C ALCMs armed with a 454-kg (1,000-lb) blast fragmentation warhead

Saab 35 Draken

The J 35J conversion programme will keep Sweden's Draken J 35F interceptors viable until well into the 1990s. It adds systems improvements and two new inboard wing pylons.

The **Saab 35 Draken** was developed to meet an ambitious 1949 requirement for a new fighter for the Swedish Flygvapen. The design adopted a distinctive double-delta configuration, and was first flown in prototype form on 25 October 1955. The initial production **J 35A** was powered by a licence-built RM6B (Avon) turbojet (featuring a more efficient Swedish afterburner) and entered service in 1960.

Following interceptor variants comprised the **J 35B** with a lengthened rear fuselage and twin retractable tailwheels; and **J 35D** with the more powerful RM6C engine (requiring larger inlets), greater fuel capacity, and more advanced radar and equipment. The final **J 35F-1** variant introduced more capable radar and collision-course fire control, a bulged canopy, only one 30-mm cannon, a new afterburner, increased fuel capacity, and licence-built Falcon AAMs. The **J 35F-II** featured an IR sensor. The J 35D-based reconnaissance **S 35E** replaced the radar nose with a new section housing five cameras. The tandem two-seat **Sk 35C** trainer version was first flown on 30 December 1959. It has no operational capability and features an instructor's periscope.

Most Drakens have now been retired from Swedish service. Sixty-four J 35Fs were converted to **J 35J** standard with two additional pylons and improvements to the radar, IR sensor, IFF, cockpit displays and avionics. A few Sk 35Cs are active with F10 to support the remaining J 35Js.

Export Drakens

The J 35F-based **Saab 35X** was developed for export with increased attack capability. Denmark ordered 46 aircraft during 1968/69, comprising **A 35XD** fighter-bombers (designated **F-35**); **RF-35** recce/fighters with S 35E camera nose; and **TF-35** (**Sk 35XD**) trainers. Danish Drakens were upgraded in the mid-1980s with a nav/attack computer, an INS, a HUD and an LRMTS in a reprofiled nose. They were retired in December 1993. Finland bought 12 new **J 35XS**s, and leased six Swedish **J 35BS** for training purposes in 1975. It later purchased 24 ex-Flyg-vapen J 35Fs (**J 35FS**) and five Sk 35Cs (**J 35CS**). Austria's 24 **J 35Ö**s are overhauled Swedish J 35Ds acquired from 1988.

The J 35A-derived two-seat Sk 35C Draken trainer features a periscope for improved forward view from the rear cockpit.

Saab J 35F Draken

Specification: Saab J 35J Draken
Powerplant: one Volvo Flygmotor RM6C turbojet (licence-built Rolls-Royce Avon Series 300) rated at 56.89 kN (12,790 lb st) dry and 78.51 kN (17,650 lb st) with afterburning
Dimensions: wing span 9.40 m (30 ft 10 in); length 15.35 m (50 ft 4 in); height 3.89 m (12 ft 9 in); wing area 49.20 m² (529.60 sq ft)
Weights: empty 8250 kg (18,188 lb); normal take-off 11400 kg (25,132 lb); maximum take-off 12270 kg (27,050 lb) in the interception role
Performance: maximum level speed 'clean' at 10975 m (36,000 ft) more than 1,147 kt (2126 km/h; 1,321 mph); maximum rate of climb at sea level 10500 m (34,450 ft) per minute with afterburning; combat radius 304 km (564 km; 350 miles) on a hi-lo-hi attack mission with internal fuel
Armament: one ADEN M/55 30-mm cannon in starboard wing root; maximum ordnance 2900 kg (6,393 lb), including Rb28 Falcon (AIM-4D IR-homing), RB27 Falcon (AIM-26B SARH-homing) and Rb24 (AIM-9P Sidewinder) AAMs, typical load is four missiles and two fuel tanks

Saab 37 Viggen

Sweden
Multi-role fighter

Maintaining Sweden's long tradition of deploying capable indigenous combat aircraft, the Viggen forms the backbone of its air power. These JA 37 interceptors serve with F13 wing.

Saab's **System 37** was developed as a relatively low-cost Mach 2 multi-role fighter capable of short-field operations. The design pioneered the use of flap-equipped canards with a stable delta-wing configuration. The selected RM8A turbofan was based on the commercial P&W JT8D-22 and equipped with a thrust reverser and Swedish afterburner.

The initial **AJ 37 Viggen** all-weather attack variant featured sophisticated nav/attack and landing systems and a large multi-role radar. The first of seven Viggen prototypes initially flew on 8 February 1967 and deliveries of the first of 109 AJ 37s began in 1971. The primary armament comprises Saab Rb 04E and licence-built AGM-65 ASMs.

Recce, maritime strike and trainer variants

Several AJ 37-based variants were developed. The **SF 37** is tasked with all-weather day and night overland reconnaissance. It is equipped with various optical and IR cameras, and carries podded sensors. The **SH 37** was modified for all-weather sea surveillance and patrol with a secondary maritime strike role. It has a modified radar, ventral night reconnaissance and long-range camera pods. A tandem two-seat **Sk 37** trainer was also developed with a stepped rear cockpit fitted with a bulged canopy and twin periscopes.

As many as 115 surviving AJ, SF and SH 37s are to be upgraded with a digital databus to multi-role **AJS 37** standard, combining their various specialised systems with some of the JAS 39's weapon systems. Modified AJS 37s started re-equipping the Flygvapen from late 1993.

Interceptor

The **JA 37** was developed as a dedicated interceptor (with a secondary ground-attack role) to replace J 35 Drakens. It introduced a new pulse-Doppler look-down/shoot-down radar, new avionics, an uprated and modified RM8B engine and a ventral 30-mm cannon. A modified AJ 37 was flown as the JA 37 prototype on 27 September 1974. Overall production of this variant was increased to 149, taking total Viggen procurement to 330. JA 37s entered service in 1978 and have equipped eight air defence *divisionen* (squadrons).

The Viggen's versatility produced several variants. Attack-tasked AJ 37s wear a distinctive 'splinter' camouflage.

Saab SF 37 and SH 37 (side view)

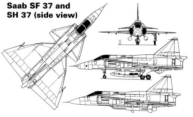

Specification: Saab JA 37 Viggen
Powerplant: one Volvo Flygmotor RM8B turbofan (P&W JT8D-22 with Swedish afterburner and thrust reverser) rated at 72.06 kN (16,200 lb st) dry and 125.04 kN (28,110 lb st) with afterburning
Dimensions: wing span 10.60 m (34 ft 9.25 in); canard foreplane span 5.45 m (17 ft 10.5 in); length 16.40 m (53 ft 9.75 in) including probe; height 5.90 m (19 ft 4.25 in); wing area 46.00 m² (495.16 sq ft); canard foreplane area 6.20 m² (66.74 sq ft)
Weights: normal take-off 15000 kg (33,069 lb); maximum take-off 17000 kg (37,478 lb) for intercept or 20500 kg (45,194 lb) for attack
Performance: maximum level speed 'clean' at 10975 m (36,000 ft) more than 1,147 kt (2126 km/h; 1,321 mph; combat radius more than 540 nm (1000 km; 622 miles) on hi-lo-hi profile
Armament: one ventral 30-mm Oerlikon KCA cannon with 150 rounds; primary armament of up to six Rb 71 (Sky Flash) and Rb 74 (AIM-9L) AAMs; maximum ordnance 13,000 lb (5897 kg) for secondary air-to-surface role, including four pods each containing six Bofors 5.3-in (13.5-cm) rockets

Saab JAS 39 Gripen

The Gripen typifies the new generation of advanced canard delta combat aircraft. This Viggen successor offers the same multi-role capability in a cheaper and lighter airframe.

Following the cancellation of the Saab 38 (B3LA) light-attack/advanced trainer project in 1979, Saab began development of the **JAS 39 Gripen** as a lightweight multi-role Viggen successor. A contract was signed for the construction of five prototypes and 30 initial production aircraft, with options for a further 110. The Gripen's **Jagt Attack Spaning** (fighter/attack/reconnaissance) designation reflects the planned roles for a single design using podded equipment and press-button programmable software changes. This design emerged as a close-coupled delta canard, powered by a single licence-built and developed RM12 (GE F404-400) engine with a new afterburner. Although it lacks a thrust reverser, it is still intended to operate from 800-m (875-yd) emergency motorway airstrips.

As a CCV, the Gripen is claimed to be the first inherently unstable canard production fighter. Other advanced technologies include a pulse-Doppler look-down multi-mode radar, a full-authority triplex digital FBW system, cockpit MFDs, wide-angle holographic HUD, HOTAS and the use of carbon-fibre composites for about 30 per cent of the airframe. Gripen weapon options include the Saab Rb 15F ASM, BK/DWS sub-munitions dispenser, advanced AAMs, and reconnaissance and ECM pods.

First flight and procurement

The prototype Gripen first flew on 9 December 1988 but crashed in February 1989 due to FBW computer software control law deficiencies, putting the programme significantly behind schedule. A further Gripen (the first production aircraft) crashed in 1993. The second production batch of 110 was finally authorised in June 1992. This procurement includes 14 two-seat **JAS 39B** operational trainers, to fly from 1996 onwards. Delivery of the 30 first-batch Gripens is due after 1994, initially to equip F7 at Såtenäs.

An upgraded **JAS 39C** has been revealed, featuring an uprated RM12, improved computers and more weapons, for both potential export (**JAS 39X**) and for a third production batch of up to 180 Gripens to replace the last of some 350 Viggens and re-equip all 16 SAF combat squadrons.

The second prototype Gripen carries three Bofors rocket pods, two wingtip Rb 74 missiles and two test cameras.

Saab JAS 39 Gripen

Specification: Saab JAS 39A Gripen
Powerplant: one Volvo Flygmotor RM12 (licence-built General Electric F404-GE-400) turbofan rated at 54.00 kN (12,140 lb st) dry and 80.51 kN (18,100 lb) with afterburning
Dimensions: wing span 8.00 m (26 ft 3 in); length 14.10 m (46 ft 3 in); height 4.70 m (15 ft 5 in)
Weights: operating empty 6622 kg (14,599 lb); normal take-off about 8000 kg (17,637 lb); maximum take-off 12473 kg (27,498 lb)
Performance: (estimated) maximum level speed 'clean' at 36,000 ft (10975 m) 1,147 kt (2126 km/h; 1,321 mph); supersonic at all altitudes
Armament: one ventral 27-mm Mauser Bk 27 cannon plus two wingtip-mounted Rb 74 (AIM-9L) or other IR-homing AAMs; maximum ordnance 6500 kg (14,330 lb), including medium-range active-radar AAMs (Active Sky Flash, S-225X project, AIM-120 and MICA), Rb 75 (Maverick) and Saab Rb 15F ASMs, DWS 39 stand-of munitions dispenser, conventional or retarded bombs, rockets, fuel tanks or FLIR, reconnaissance and EW pods

SEPECAT Jaguar

Single-seat strike fighter and advanced trainer

India's sizeable Jaguar fleet includes 15 two-seat Jaguar IT operational trainers. Ten have been assembled under licence.

Produced to meet a 1965 joint Anglo-French specification for an advanced trainer/tactical support aircraft, the **SEPECAT Jaguar** was transformed into a potent fighter-bomber with·sophisticated nav/attack systems for the low-level all-weather attack role. The first of eight Jaguar prototypes made its maiden flight on 8 September 1968.

The RAF received 200 Jaguars, comprising 165 single-seat **GR.Mk 1**s (with chisel laser noses) and 35 **T.Mk 2** trainers. The latter have the full nav/attack avionics suite but no lasers. Delivered in 1973-78, GR.Mk 1s were tasked with nuclear strike, reconnaissance and conventional strike. Recce aircraft carry a centreline pod containing five cameras and an IR linescan. Only the Coltishall Wing remains, its GR.Mk 1s gaining Adour Mk 104 engines from 1978-84. The **GR.Mk 1A** upgrade added a FIN1064 INS, AN/ALE-40 flare dispensers, Phimat flare pods and jamming pods to 75 single-seaters. Fourteen trainers were similarly upgraded as **T.Mk 2A**s. For Operation Granby in 1991 Jaguars also used CRV-7 rockets and CBU-87 cluster bombs.

French Jaguars

The Armée de l'Air's 160 single-seat **Jaguar A**s have generally less capable avionics than RAF Jaguars, but remain effective strike aircraft. Various systems have been added including TAV-38 or ATLIS laser designators and the OMERA 40 camera. The 40 cannon-armed **Jaguar E** trainers lack a full nav/attack avionics fit. Initial deliveries were made from January 1972, and some Jaguar As were tasked with pre-strategic nuclear-strike with AN 52 weapons (withdrawn in 1991). The surviving 85 Jaguar As now undertake tactical support/ground-attack missions. AA Jaguars have seen action in Mauritania, Chad and the Gulf.

Jaguar International

All export **Jaguar International**s are based on the RAF's **Jaguar B/S** airframe. Customers comprise Ecuador, India, Nigeria and Oman. India is the biggest Jaguar operator today, with **Jaguar IS** strike, **IT** trainer and **IM** maritime strike aircraft. The latter have Agave radar in a reprofiled nose and are armed with BAe Sea Eagle missiles.

A French Jaguar A sets forth on an Opération Daguet sortie, carrying an AS30L ASM, R550 AAM and an ATLIS laser pod.

SEPECAT Jaguar international IM

Specification: SEPECAT Jaguar A
Powerplant: two Rolls-Royce/Turboméca Adour Mk 102 turbofans each rated at 22.75 kN (5,115 lb st) dry and 32.49 kN (7,305 lb st) with afterburning
Dimensions: wing span 8.69 m (28 ft 6 in); length 17.53 m (57 ft 6.25 in) including probe; height 4.89 m (16 ft 0.5 in); wing area 24.18 m² (260.27 sq ft)
Weights: empty equipped 7000 kg (15,432 lb); normal take-off 10954 kg (24,149 lb); maximum take-off 15700 kg (34,612 lb)
Performance: maximum level speed 'clean' at 10975 m (36,000 ft) 917 kt (1699 km/h; 1,056 mph); combat radius 460 nm (852 km; 530 miles) on a hi-lo-hi attack mission with internal fuel
Armament: two DEFA 553 30-mm cannon with 150 rpg; maximum ordnance 4536 kg (10,000 lb) including laser-guided BGL 1,000-lb (454-kg) bombs, AS30L ASMs, AS37 Martel ARMs; 125-, 250- and 400-kg bombs; Belouga cluster bombs; 36 x 68 mm F1 and 4 x 100 mm R3 rocket pods; BAP-100 anti-runway and BAT-120 area-denial bomblets; and R550 Magic 2 self-defence AAMs

Shenyang J-8 'Finback'

People's Republic of China
Single-seat interceptor

The demise of the US 'Peace Pearl' avionics upgrade, and China's purchase of Su-27 'Flankers', have probably sealed the fate of the indigenously-developed J-8II 'Finback'.

The **J-8** originated from a 1964 PLA requirement for a fighter with performance superior to that of the MiG-21. Shenyang's twin-engined design incorporated a scaled-up configuration of the MiG-21's 'tailed delta', with a ranging radar in the intake centrebody. The first of two prototypes flew on 5 July 1969. These undertook a protracted flight test programme (lasting 10 years), which was interrupted largely for political reasons. The J-8 was armed with a single 30-mm cannon and up to four underwing PL-2 AAMs.

The improved **J-8I** was designed as an all-weather fighter, and featured a new Sichuan SR-4 radar in an enlarged intake centrebody, some aerodynamic refinements and a two-piece canopy. It introduced a twin-barrelled 23-mm 23-III cannon, and provision for four rocket pods as an alternative to the AAMs. A prototype flew on 24 April 1981. J-8 and J-8I production totalled about 100 aircraft.

J-8II 'Finback-B'

Development of the further improved **J-8II 'Finback-B'** began in May 1981, leading to first flight of a prototype on 12 June 1984. The J-8II introduced relocated new lateral air intakes for its 69-kN (15,430-lb st) WP-13B turbojets and a ventral folding fin. Small-scale batch production of the J-8II was undertaken, but the type may not have entered service. An export **F-8B** version introduced a pulse-Doppler look-down radar and digital avionics, with a HUD and two HDDs. It has not won any orders.

On 5 August 1987 Grumman received a contract to design, develop and test an avionics upgrade for the J-8II. This introduced a modified AN/APG-66 radar, giving compatability with BVR SARH missiles like the AIM-7 Sparrow. The aircraft would also receive a modern HUD, a US ejection seat and an INS, and a bubble canopy and frameless wraparound windscreen. However, the Tienanmen Square massacre led to an immediate halt on work on the project. Development restarted, but China withdrew from the project, leading to its cancellation. Coupled with the delivery of Su-27 'Flankers' to China, this may have killed off the J-8.

China's first indigenous fighter underwent a protracted development. The initial J-8I variant (note the small intake-mounted ranging radar) was mainly used for test duties.

Shenyang J-8II/F-8B 'Finback-B'

Specification: Shenyang J-8 II/F-8B 'Finback-B'
Powerplant: two Liyang (LMC) Wopen-13A II turbojets each rated at 42.66 kN (9,590 lb st) dry and 65.90 kN (14,815 lb st) with afterburning
Dimensions: wing span 9.34 m (30 ft 8 in); length 21.59 m (70 ft 10 in) including probe; height 5.41 m (17 ft 9 in); wing area 42.20 m² (454.25 sq ft)
Weights: empty 9820 kg (21,649 lb); normal take-off 14300 kg (31,526 lb); maximum take-off 17800 kg (39,242 lb)
Performance: maximum level speed 'clean' at 10975 m (36,000 ft) 1,262 kt (2338 km/h; 1,453 mph); maximum rate of climb at sea level 12000 m (39,370 ft) per minute; service ceiling 20200 m (66,275 ft); ferry range 1,187 nm (2200 km; 1,367 miles) with drop tanks; combat radius 432 nm (800 km; 497 miles)
Armament: one ventral Type 23-3 twin-barrelled 23-mm cannon with 200 rounds; ordnance includes PL-2B IR-homing and PL-10 medium-range SARH AAMs, HF-16B 12-round pod containing 57-mm air-to-air rockets, bombs and fuel tanks

Sikorsky S-61 Sea King

Agusta's AS-61R Pelican is a licence-built Sikorsky HH-3F, and is deployed around the Italian coastline for SAR duties.

Sikorsky developed the **S-61 Sea King** to replace its previous S-58 design, combining dual ASW hunter/killer roles in a single airframe. The prototype **YHSS-2** first flew on 11 March 1959 and was followed by 245 production **SH-3A**s for the USN. The primary sensors were an AQS-10 dipping sonar and an APN-130 search radar; in the 'killer' role the SH-3A carried two torpedoes or depth charges.

The following 74 **SH-3D**s introduced uprated T58-GE-10 engines, AQS-13A sonar and APN-182 radar. The **SH-3G** conversion modified 105 SH-3A/Ds to act as general-purpose rescue platforms and transports. Few SH-3Gs remain in service. The **SH-3H** standard involved 116 SH-3A/-D/-Gs converted to perform the inner-zone ASW mission with AQS-13B sonar, LN66HP radar, chaff dispensers and an ASQ-81 towed MAD bird. ESM equipment and the radar were later replaced by a modern tactical navigation suite and improved sonobuoy and sonar processing capability.

By 1994 eight USN SH-3H units were operational, along with small numbers of **UH-3A** and **VH-3A** utility transports.

Licence-production and exports

The S-61 was exported to Argentina, Brazil, Denmark, Malaysia, Peru and Spain. Licence manufacture was undertaken by Canada, Italy and Japan. Canada's 32 surviving **CH-124**s have been updated for continued ASW service. The JMSDF's Mitsubishi-built **HSS-2/-2A** and **-2B** helicopters are operated on ASW and SAR tasks. As the **AS-61A-4**, Agusta built an SH-3D logistic/VIP transport derivative for Italy, Iraq, Iran, Egypt and the Saudi Arabia. The SAR **AS-61R Pelican** was built for Italy and Venezuela.

USAF 'Jolly Green Giants' and USCG Pelicans

The USAF's **S-61R** or (**CH-3C**) long-range transport has a redesigned rear fuselage (with a rear loading ramp), large sponsons and nosewheel undercarriage. The improved **CH-3E** featured T58-GE-5 engines. **HH-3E** 'Jolly Green Giants' served as Vietnam rescue platforms, equipped with IFR probes. The USCG's 40 **HH-3F Pelican** long-range SAR lack the HH-3E's combat-related equipment, but have a nose search radar, and can carry up to 15 stretchers.

A US Navy SH-3A of HS-6 trails its MAD gear during off-shore training operations.

Sikorsky S-61
(SH-3H Sea King)

Specification: Sikorsky S-61 (SH-3H Sea King)
Powerplant: two General Electric T58-GE-10 turboshafts each rated at 1044 kW (1,400 shp)
Dimensions: main rotor diameter 18.90 m (62 ft 0 in); length overall, rotors turning 22.15 m (72 ft 8 in); fuselage 16.69 m (54 ft 9 in); height overall 5.13 m (16 ft 10 in); main rotor disc area 280.47 m² (3,019.0 sq ft)
Weights: empty 5601 kg (12,350 lb); maximum take-off 9526 kg (21,000 lb)
Performance: maximum level speed 'clean' at optimum altitude 144 kt (267 km/h; 166 mph); economical cruising speed 118 kt (219 km/h; 136 mph); maximum rate of climb at sea level 670 m (2,200 ft) per minute; service ceiling 4480 m (14,700 ft); hovering ceiling 3200 m (10,500 ft) in ground effect and 2500 m (8,200 ft) out of ground effect; range 542 nm (1005 km; 625 miles)
Armament: maximum ordnance 381 kg (840 lb)

Sikorsky S-65/-80/CH-53

The US Navy's MH-53E combines the CH-53E's airframe and three-engined installation with the RH-53D's mine-sweeping equipment. The grossly enlarged sponsons house fuel.

The **Sikorsky S-65/CH-53** heavy-lift helicopter first flew in prototype form on 14 October 1964 and, as the **CH-53A Sea Stallion**, it became the USMC's principal Vietnam heavy-lift type, entering service in 1965. The following **CH-53D** introduced uprated engines and a revised interior for increased troop accommodation. In the air assault role, it can carry 55 troops or 3630 kg (8,000 lb) of cargo internally. Seven USMC units operate the CH-53D on supplies and equipment transportation duties. S-65s were exported to Austria, West Germany (112 **CH-53G**s) and Israel (whose **CH-53 2000** upgrade adds new EW and cockpit systems).

Three-engined variant
The improved **S-80/CH-53E Super Stallion** variant flew in prototype form in 1974, introducing a third engine, a seven-bladed main rotor, a lengthened airframe and fuselage sponsons. IOC was achieved in 1981 and the type now forms a vital part of USMC amphibious operations. Envisaged procurement calls for 177 aircraft.

Mine sweeping variants were developed to tow a mine countermeasures 'sled'. The definitive **RH-53D** introduced uprated engines and optional IFR probe and sponson tanks. The current CH-53E-based **MH-53E Sea Dragon** features grossly enlarged fuselage sponsons with increased fuel capacity and modernised mine countermeasures systems. The JMSDF operates 11 similar **S-80M-1**s.

USAF combat rescue and special operations models
The USAF operated **HH-53B/-53C** rescue platforms with IFR probe and external tanks, and **CH-53C** trainer/transports. The **HH-53H Pave Low III**s had LLLTV, APQ-158 TF radar, Doppler and INS, a FLIR turret, a map display system and numerous countermeasures. The Special Operations **MH-53H** has NVG-compatibility. The current **MH-53J Pave Low III Enhanced** has uprated engines, TFR, FLIR, NVG, RWRs, IR jammers, chaff/flare dispensers, GPS, IFR probe, external tanks, titanium armour and provision for three door/rear ramp-mounted 7.62-mm Miniguns. It is tasked primarily with Special Forces support.

The CH-53A quickly proved its worth during the Vietnam War. One USMC squadron recovered nearly 1,100 aircraft.

Sikorsky MH-53J Pave Low Enhanced

Specification: Sikorsky S-80 (CH-53E Super Stallion)
Powerplant: three General Electric T64-GE-416 turboshafts each rated at 3266 kW (4,380 shp) for 10 minutes, 3091 kW (4,145 shp) for 30 minutes and 2756 kW (3,696 shp) for continuous running
Dimensions: main rotor diameter 24.08 m (79 ft 0 in); length overall, rotors turning 30.19 m (99 ft 0.5 in), fuselage 22.35 m (73 ft 4 in); height overall, rotors turning 8.97 m (29 ft 5 in); main rotor disc area 455.38 m² (4,901.67 sq ft)
Weights: empty 15072 kg (33,338 lb); maximum take-off 31640 kg (69,750 lb) with an internal payload or 33340 kg (73,500 lb) with an external payload; maximum payload 16330 kg (36,000 lb)
Performance: maximum level speed 'clean' at sea level 170 kt (315 km/h; 196 mph); cruising speed at sea level 150 kt (278 km/h; 173 mph); maximum rate of climb at sea level with 11340-kg (25,000-lb) payload 762 m (2,500 ft) per minute; service ceiling 5640 m (18,500 ft); operational radius 500 nm (925 km; 575 miles) with 9072-kg (20,000-lb) external payload

Sikorsky UH-60 Black Hawk

The S-70/UH-60 family has achieved widespread export sales. Australia received 16 S-70B-2 Seahawks and 39 S-70A-9s, at first assigned to the RAAF, but later turned over to army.

Sikorsky MH-60G Pave Hawk

Sikorsky's **S-70** design was developed to meet the US Army's 1972 requirement for a utility/tactical transport helicopter to replace the Bell UH-1, offering far better performance, crashworthiness and all-round survivability. The **YUH-60A** prototype first flew on 17 October 1974 and featured a broad, squat cabin for one crew chief/door gunner and 11 troops. A production **UH-60A Black Hawk** first flew in 1978, and the type entered service in June 1979. The cheaper **UH-60L** (delivered from 1989) has uprated engines and can lift an HMMWV vehicle with TOW anti-tank installation. The Army hopes to procure at least 1,400 UH-60A/Ls, and it may also acquire hoist-equipped **UH-60Q**s for medical evacuation ('Dustoff') duties. The export **S-70A** is in widespread service worldwide. Licence-production is undertaken in Korea (**UH-60P**) and Japan (**UH-60J**).

Special missions variants

The special electronics mission **EH-60A** was tested with Quick Fix IIB battlefield jamming system. Production **EH-60C**s have two tailboom dipole antennas, and a ventral retractable whip aerial. The **MH-60A** Special Operations support helicopter has FLIR, extra nav/comms, auxiliary fuel tanks and Miniguns. The following **MH-60L** 'Velcro Hawks' have matching systems. The definitive **MH-60K** has TFR, FLIR, pintle-mounted 0.50-in machine-guns, ESSS wings for fuel tanks, IFR probe, HIRSS, comprehensive comms/nav equipment, and defensive warning receivers and countermeasures.

Combat rescue Hawks

The cancelled **HH-60D/E** USAF combat rescue models were replaced by the **HH-60A**, pitched halfway in terms of avionics fit. The USAF pursued a three-phase procurement process; the Phase One **UH-60A Credible Hawk** has an IFR probe, provision for additonal fuel and door guns. Sixteen Special Operations **MH-60G** support helicopters have 0.50-in machine-guns, Phase Two avionics (radar, GPS, INS, secure comms equipment and full countermeasures) and Phase Three equipment (HUD and FLIR). **HH-60G**s have Phase Two avionics and lower calibre door weapons.

This USAF UH-60A is seen after conversion to Credible Hawk configuration. The IFR probe extends ahead of the rotor disc.

Specification: Sikorsky S-70A (UH-60A Black Hawk)
Powerplant: two General Electric T700-GE-700 turboshafts each rated at 1151 kW (1,560 shp)
Dimensions: main rotor diameter 16.36 m (53 ft 8 in); length overall, rotors turning 19.76 m (64 ft 10 in), fuselage 15.26 m (50 ft 0.75 in); height overall 5.13 m (16 ft 10 in); stabiliser span 4.38 m (14 ft 4.5 in); main rotor disc area 210.14 m^2 (2,262.03 sq ft)
Weights: empty 5118 kg (11,284 lb); normal take-off 7708 kg (16,994 lb); maximum take-off 9185 kg (20,250 lb); maximum internal payload 1197 kg (2,640 lb) and 3629 kg (8,000 lb) carried externally
Performance: maximum level speed 'clean' at sea level 160 kt (296 km/h; 184 mph); maximum cruising speed at 1220 m (4,000 ft) 145 kt (268 km/h; 167 mph); maximum vertical rate of climb at 1220 m (4,000 ft) 125 m (411 ft) per minute; service ceiling 5790 m (19,000 ft); range 319 nm (592 km; 368 miles) with standard fuel; endurance 2 hours 18 minutes

Sikorsky SH-60 Seahawk

The SH-60F Ocean Hawk is optimised for inner-zone ASW protection of the carrier battle group, complementing the fixed-wing S-3 Viking, which deals with longer-range threats.

A navalised H-60 was developed by Sikorsky to meet the USN's LAMPS III requirement. This called for a helicopter capable of providing an over-the-horizon search and strike capability for ASW frigates and destroyers. A prototype **YSH-60B** flew on 12 December 1979, followed by the first production **SH-60B Seahawk** in 1983. Retaining some 83 per cent commonality with the UH-60A, it introduced airframe anti-corrosion treatment, T700-GE-401 engines, and RAST (recovery assist secure and traverse) gear. Mission equipment comprises APS-124 ventral search radar, a 25-tube sonobuoy launcher, ASQ-81(V)2 towed MAD, ALQ-142 ESM antennas, datalinks, secure comms and various on-board computers and processing units. The SH-60B undertakes final localisation of submarines (using MAD) and prosecutes attacks using two Mk 46 torpedoes. It has a hoist for a secondary SAR role and can also perform utility transport missions. The USN intends to procure 260 SH-60Bs. They have been exported to Australia, Greece, Japan and Spain.

A second ASW **SH-60F Ocean Hawk** variant performs the CVW Inner Zone ASW mission. Initially deployed in 1991, it deletes LAMPS and RAST equipment in favour of an AQS-13F dipping sonar and improved detection systems such as FLIR and ESM, with options for MAD and search radar. The SH-60F is armed with three Mk 50 torpedoes. Taiwan has bought 10 similar **S-70C(M)-1 Thunderhawk**s.

Rescue platform

The SH-60F-based **HH-60H Rescue Hawk** is the USN's strike rescue platform, entering service in 1989. It has a secondary covert role of infil/exfil of SEAL commando teams. All HH-60Hs have HIRSS exhaust suppression and 7.62-mm (later GECAL 0.50-in) door guns. Following Desert Storm experience, they will acquire turreted FLIR, lengthened sponsons for more defensive systems, and forward-firing armament (2.75-in rocket pods and 0.50-in guns).

The USCG's SAR **HH-60J Jayhawk** replaces the HH-3F and has a search/weather radar, a searchlight, an NVG-compatible cockpit and an optional fuel tank. The HH-60J has a secondary drug interdiction role.

The SH-60B Seahawk is the US Navy's standard shipborne ASW helicopter, flying from frigates and destroyers.

Sikorsky SH-60J

Specification: Sikorsky S-70B (SH-60B Seahawk – aircraft delivered from 1988)
Powerplant: two General Electric T700-GE-401C turboshafts each rated at 1417 kW (1,900 shp)
Dimensions: main rotor diameter 16.36 m (53 ft 8 in); length overall, rotors turning 19.76 m (64 ft 10 in), fuselage 15.26 m (50 ft 0.75 in); height overall, rotor turning 5.18 m (17 ft 0 in); stabiliser span 4.38 m (14 ft 4.5 in); main rotor disc area 210.05 m² (2,262.03 sq ft)
Weights: empty 6191 kg (13,648 lb) for the ASW mission; ASW mission take-off 9182 kg (20,244 lb) or 8334 kg (18,373 lb) for the ASST mission; maximum take-off 9926 kg (21,884 lb) for the utility mission; maximum payload 3629 kg (8,000 lb)
Performance: dash speed at 1525 m (5,000 ft) 126 kt (234 km/h; 145 mph); maximum vertical rate of climb at sea level 213 m (700 ft) per minute; operational radius 50 nm (92.5 km; 57.5 miles) for a 3-hour loiter, or 150 nm (278 km; 173 miles) for a 1-hour loiter
Armament: one AGM-119B Mod 7 Penguin ASM, and pintle-mounted 0.50-in machine-guns

Sukhoi Su-17 'Fitter'

Russia
Fighter-bomber

The Su-22M-4 was the final major 'Fitter' variant, identified by the deep forward fuselage and fin extension.

To improve the range and STOL capability of the Su-7, Sukhoi produced a derivative with a VG wing. The resulting **S-22I** (**Su-7IG 'Fitter-B'**) flew on 2 August 1966 and was followed by pre-production **Su-17s**. The production **Su-17M 'Fitter-C'** introduced the 109.83-kN (24,690-lb st) AL-21F-3 engine and a new nav/attack system. 'Fitter-Cs' remain in front-line service in Poland and, under the export designation **Su-20,** were delivered to Afghanistan, Algeria, Angola, Egypt, Iraq, North Korea, Syria and Vietnam. A handful of 'Fitter-Cs' were built for reconnaissance duties, with provision for multi-sensor reconaissance pods as **Su-17Rs**.

The **Su-17M-2D 'Fitter-D'** introduced a lengthened, drooping nose and revised avionics. The ranging radar was replaced by a laser ranger and Doppler was added below the nose. A sanitised version was built for export under the designation **Su-17M-2K**. This was powered by a Tumanskii R-29BS-300 and was operated by Angola, Libya and Peru.

Su-17M-3 and Su-17M-4

The **Su-17M-3** had a deepened forward fuselage, tall tailfin and removable ventral fin, like the trainer, but with two wingroot cannon and a single cockpit. A dedicated AAM launch rail was added beneath each inner wing. It was delivered to Frontal Aviation, and the similar **Su-22M-3K** (with the Tumanskii R-29BS engine) was exported to Angola, Hungary, Libya, Peru and both Yemens. The **Su-22M-4 'Fitter-K'** introduced new avionics and compatibility with an even wider range of weapons, and is in widespread service with a number of CIS air forces. It was exported to Poland, Czechoslovakia, East Germany and Afghanistan.

The first trainer was the **Su-17UM-2D 'Fitter-E'**, and retained all avionics (including the laser rangefinder), but had no port cannon. Export **Su-17UM-2Ks** used the R-29BS-300 engines. The next two-seater was based on the tall-tailed Su-17M-3 airframe and as the **Su-22UM-3K** was produced with Lyul'ka and Tumanskii engines. The **'Fitter-G'** is by far the fastest Su-17, capable of reaching Mach 2.1, while single-seaters are limited to Mach 1.7.

Libya's fleet of 'Fitters' comprises mostly M-2Ks and M-3Ks. Some have been used as interceptors, armed with AA-2 'Atoll' AAMs. Two were shot down by USN F-14s in 1981.

Sukhoi Su-17M-3

Specification: Sukhoi Su-17M-4 'Fitter-K'
Powerplant: one NPO Saturn (Lyul'ka) AL-21F-3 turbojet rated at 76.49 kN (17,196 lb st) dry and 110.32 kN (24,802 lb st) with afterburning
Dimensions: wing span 13.80 m (45 ft 3 in) spread and 10.00 m (32 ft 10 in); length 18.75 m (61 ft 6.25 in) including probes; height 5.00 m (16 ft 5 in); wing area about 40.00 m² (430.57 sq ft) spread and about 37.00 m² (398.28 sq ft) swept
Weights: normal take-off 16400 kg (36,155 lb); maximum take-off 19500 kg (42,989 lb)
Performance: maximum level speed 'clean' at sea level 756 kt (1400 km/h; 870 mph); service ceiling 15200 m (49,870 ft); combat radius 621 nm (1150 km; 715 miles) on a hi-lo-hi attack mission with a 2000-kg (4,409-lb) warload
Armament: two wingroot-mounted NR-30 30-mm cannon with 80 rpg; maximum practical ordnance load 1000 kg (2,205 lb) when drop tanks are carried, including a wide variety of free-fall bombs and podded and unpodded unguided rockets

Sukhoi Su-24 'Fencer'

Russia's main low-level strike platform, the improved Su-24M 'Fencer-D' variant, has a reshaped nose radome housing attack and terrain-following radars.

The **Su-24** was intended as an all-weather low-level supersonic bomber able to attack fixed and mobile targets with pinpoint accuracy and with a secondary photographic reconnaissance role. Sukhoi initially developed the **T-6-1** delta-winged VTOL bomber, with separate cruise and lift engines. It proved unsuccessful, and a VG wing was fitted to the **T-6-2IG** prototype, removing the heavy lift jets to leave space for fuel or weapons. The aircraft made its maiden flight during May 1970.

The production Su-24 **'Fencer-A'** was powered by a pair of Perm/Soloviev AL-21F-3 turbofans. **'Fencer-B'** had a rear fuselage more closely following the jet pipes, and introduced a brake chute fairing below the rudder. **'Fencer-C'** has triangular RWR fairings on the sides of the fin-tip and on the engine intakes. The aircraft could carry the free-fall TN-1000 and TN-1200 nuclear bombs, and a variety of conventional free-fall bombs and guided ASMs. 'Fencer-Bs' and 'Fencer-Cs' remain in widespread front-line use with Russia, and with a number of former Soviet states.

'Fencer-D', reconnaissance and Elint variants

The improved **Su-24M 'Fencer-D'** attack variant entered service in 1986 and introduced a retractable IFR probe above the nose, an upgraded avionics suite and provision for a UPAZ-A buddy refuelling pod. Its shortened, reshaped radome houses Orion-A forward-looking attack radar and Relief TFR. The Kaira 24 laser and TV sighting system gives compatibility with the newest Soviet TV- and laser-guided ASMs. Export **Su-24MK**s have been delivered to Iran, Iraq (all 24 of the latter being absorbed into the Iranian air force during Desert Storm), Libya and Syria.

The **Su-24MR 'Fencer-E'** tactical reconnaissance aircraft uses internal and podded sensors of various types, and is able to transmit data from some sensors to a ground station in real time. The **Su-24MP 'Fencer-F'** is believed to have a primary Elint-gathering role and is similar to the Su-24MR. It can be distinguished from the earlier aircraft by a prominent undernose fairing below the nose and swept-back intake-mounted 'hockey stick' antennas.

The Su-24 is a formidable warplane of similar capabilities to the F-111. It is also used for reconnaissance missions.

Sukhoi Su-24M 'Fencer-D'

Specification: Sukhoi Su-24 'Fencer-C'
Powerplant: two NPO Saturn (Lyul'ka) AL-21F-3A turbofans each rated at 76.49 kN (17,196 lb st) dry and 110.32 kN (24,802 lb st) with afterburning thrust
Dimensions: wing span 17.63 m (57 ft 10 in) spread and 10.36 m (34 ft 0 in) swept; length 24.53 m (80 ft 5.75 in) including probe; height 4.97 m (16 ft 3.75 in); wing area 42.00 m² (452.10 sq ft)
Weights: empty equipped 19000 kg (41,887 lb); normal take-off 36000 kg (79,365 lb); maximum take-off 39700 kg (87,522 lb)
Performance: maximum level speed 'clean' at 11000 m (36,090 ft) 1,251 kt (2320 km/h; 1,441 mph); service ceiling 17500 m (57,415 ft); combat radius 1050 km (565 nm; 650 miles) on a hi-lo-hi attack mission with a 3000-kg (6,614-lb) warload and two drop tanks
Armament: one GSh-6-23M 23-mm cannon; maximum ordnance 8000 kg (17,637 lb) including laser and TV-guided ASMs, TN-1000 and TN-1200 nuclear weapons and 100-kg FAB-100 bombs

Sukhoi Su-25 'Frogfoot'

The first Warsaw Pact unit to operate the 'Frogfoot' was the Czechoslovak 'Ostrava' Regiment, based at Pardubice with 35 Su-25Ks. This Su-25K displays its split wingtip speed brakes.

The **Su-25** was developed during the late 1960s, as a jet *shturmovik*. A prototype flew on 22 February 1975, with the Su-17M2's weapons system. Further prototypes introduced the R-95Sh engine (a non-afterburning version of the MiG-21's R-13-300 turbojet), a twin-barrelled 30-mm cannon and the Su-17M3's weapons system. The production **Su-25** introduced enlarged engine intakes and increased armour around cockpit and critical components. Combat experience in Afghanistan led to the addition of bolt-on chaff/flare dispensers, and an exhaust IR suppressor. From 1987 the R-195 engine was introduced, as fitted to all two-seaters. Su-25 production ended in 1989, after 330 aircraft had been delivered, including export **Su-25K**s to Bulgaria, Czechoslovakia, Iraq (Gulf War survivors fleeing to Iran) and North Korea. The **Su-25BM** is a target tug conversion.

'Frogfoot' trainer
The **Su-25UB** combat capable trainer features a taller tailfin, and a lengthened fuselage with stepped cockpits. Similar **Su-25UBK**s were exported to Bulgaria, Czechoslovakia, Iraq and North Korea. The **Su-25UT** (later redesignated **Su-28**) had all armament and weapons systems removed and was intended for the pilot training role with the VVS and DOSAAF. The **Su-25UTG** is a dedicated carrier trainer with strengthened undercarriage and arrester hook.

Su-34 (Su-25T) and derivatives
The **Su-34** (originally **Su-25T**) is an extensively modernised Su-25UB derivative, using the rear cockpit to house avionics and fuel. To give true night capability, the aircraft has new avionics, with INAS and two digital computers. The nose houses a TV camera and laser designator, spot tracker and rangefinder. An LLLTV/FLIR system or a low-light level navigation system can be carried. The export **Su-25TK** has been offered to Abu Dhabi and Bulgaria.

A proposed maritime **Su-25TP** variant combines the Su-25UTG's features with the Su-25T's weapons system. A land-based Su-34 variant (redesignated **Su-25TM**) has been offered to Georgia, with podded MMW radar and a FLIR.

The Su-25UB is a combat-capable two-seat conversion trainer, serving with most 'Frogfoot' operators.

Sukhoi Su-25

Specification: Sukhoi Su-25K 'Frogfoot-A'
Powerplant: two MNPK 'Soyuz' (Tumanskii) R-195 turbojets each rated at 44.13 kN (9,921 lb st) dry
Dimensions: wing span 14.36 m (47 ft 1.4 in); length 15.53 m (50 ft 11.5 in); height 4.80 m (15 ft 9 in); wing area 33.70 m² (362.76 sq ft)
Weights: empty equipped 9500 kg (20,944 lb); normal take-off 14600 kg (32,187 lb); maximum take-off 17600 kg (38,801 lb)
Performance: maximum level speed 'clean' at sea level 526 kt (975 km/h; 606 mph); service ceiling 7000 m (22,965 ft); combat radius 297 nm (550 km; 342 miles) on hi-lo-hi attack mission with a 4000-kg (8,818-lb) warload and two drop tanks
Armament: one 30-mm AO-17A cannon in port lower fuselage, with 250 rounds; maximum ordnance 4400 kg (9,700 lb) including 57-mm to 330-mm unguided rockets, various free-fall and laser-guided bombs, cluster bombs, dispenser weapons and incendiary weapons and cannon pods; Kh-23 (AS-7 'Kerry'), Kh-25 (AS-10 'Karen') and Kh-29 (AS-14 'Kedge') ASMs, outermost hardpoint for carriage of R-60 (AA-8 'Aphid') AAMs for self-defence

Sukhoi Su-27 'Flanker'

A single-seat 'Flanker' in company with an Su-27UB combat trainer. The highly-capable Su-27 is widely regarded as the world's best fighter in terms of performance.

Sukhoi began work on a new PVO interceptor in 1969. This was to be highly manoeuvrable, with long range, heavy armament and modern sensors, and was to be capable of intercepting low-flying or high-level bombers, and of out-fighting the F-15. The **T-10** prototype flew on 20 May 1977 and was dubbed **'Flanker-A'**. Testing revealed serious problems and T-10 underwent a total redesign. The seventh protoype flew as the **T-10S-1** on 20 April 1981 with a redesigned wing, undercarriage and fuselage, and a spine-mounted airbrake, prompting the new NATO reporting name **'Flanker-B'**. A stripped **P-42** variant with uprated engines set a series of time-to-height and other records.

The Su-27 entered service during the mid-1980s, delayed by radar problems. The standard Su-27 fighter is now in service with the Russian and Ukrainian air forces, serving with Frontal Aviation and the PVO air defence force. Twenty-four **Su-27SK**s have been delivered to the People's Republic of China.

Advanced systems

The 'Flanker-B' has an advanced pulse-Doppler radar, backed up by a sophisticated EO complex with an IRST system and a laser rangefinder. This allows the Su-27 to detect, track and engage a target without using radar. The Su-27 is also compatible with a helmet-mounted target designation system. The prototype **Su-27UB** trainer flew on 7 March 1985. It featured a lengthened fuselage with stepped tandem cockpits under a single canopy and increased height/area tailfins and airbrake. Su-27UBs have been used to test buddy refuelling pods and vectoring nozzles and have spawned a new two-seat operational variant.

The **Su-27PU** (later **Su-30**) with a retractable IFR probe and offset IRST 'ball' was designed as a long-range, high-endurance interceptor. With the addition of extra equipment, the Su-30 can operate as a mini-AWACS. The single-seat **Su-27P** has the same systems as the Su-30, but its single cockpit would make long endurance difficult to exploit operationally. The **SU-30MK** is an export Su-30 derivative for multi-role use, with ground-attack capability.

The Su-27UB is the basis for the Su-30, which can be optimised for either the intercept or ground-attack roles.

Sukhoi Su-27 'Flanker-B'

Specification: Sukhoi Su-27 'Flanker-B'
Powerplant: two NPO Saturn (Lyul'ka) AL-31F turbofans each rated at 79.43 kN (17,857 lb) dry and 122.58 kN (27,557 lb st) with afterburning
Dimensions: wing span 14.70 m (48 ft 2.75 in); length 21.935 m (71 ft 11.5 in) excluding probe; height 5.932 m (19 ft 5.5 in); wing area 46.50 m^2 (500.54 sq ft)
Weights: empty 17700 kg (39,021 lb); normal take-off 22000 kg (45,801 lb); maximum take-off 30000 kg (66,138 lb)
Performance: maximum level speed 'clean' at 11000 m (36,090 ft) 1,349 kt (2500 km/h; 1,553 mph) and at sea level 729 kt (1350 km/h; 839 miles); maximum rate of climb at sea level 18300 m (60,039 ft) per minute; service ceiling 18000 m (59,055 ft); combat radius 809 nm (1500 km; 932 miles) on a hi-hi-hi interception mission with four AAMs
Armament: one 30-mm GSh-30-1 cannon in starboard wingroot with 150 rounds; maximum ordnance 6000 kg (13,228 lb) including up to six medium-range R-27 (AA-10 'Alamo') and four short-range R-73 (AA-11 'Archer') AAMs

Sukhoi Su-27/34/35

The Su-35 advanced 'Flanker' derivative is identified by much larger fins and canard foreplanes. The slightly reprofiled nose radome houses a new, longer-ranged radar.

Development of an advanced Su-27 began during the early 1980s. The proof-of-concept **T-10-24** first flew during 1982, equipped with canards. These improved agility and gave better approach characteristics and were used on the navalised **Su-27K**. This also featured folding wings and tailplanes, strengthened undercarriage and an arrester hook, but retained the original FBW control system and weapons system of the basic Su-27. The production **Su-33** would probably be based on the Su-35 FCS and radar.

Su-35 (Su-27M)

The definitive advanced Su-27 is designated **Su-35** (initially **Su-27M**) and has a completely new FCS. As well as canards, it has a new, square-topped tailfin (with internal fuel tanks). The Su-27M also has a new N-011 radar with a range of up to 400 km (or 200 km against ground targets) which can simultaneously track more than 15 targets, engaging six. The new EO complex gives compatability with advanced 'smart' weapons. An advanced datalink allows co-ordinated group operation and the tailcone houses the antenna for a rear-facing radar which will allow 'over-the-shoulder' missile shots. The Su-35 will be compatible with the new 400-km Novator KS-172 AAM-L missile. Flight testing is now reportedly complete and production has been funded.

Su-27IB

Initially designated **Su-27KU**, the **Su-27IB** combined canard foreplanes with a new forward fuselage accommodating a side-by-side two-seat cockpit. The aircraft's lack of arrester hook, slotted flaps, wing/tailplane folding and longer wheelbase mitigated against naval use, and indicated that the aircraft was the demonstrator for an Su-24 replacement. Recent articles claim a titanium armoured cockpit, armoured glass, and CRT displays. The intakes have been redesigned for higher low-level speeds. A productionised **Su-34** derivative (with suitable sensors) was flown in late 1993. A retractable IFR probe is provided on the Su-27IB, Su-34, Su-27K and Su-35. Other improvements applicable include vectoring engine nozzles.

The Su-27IB is the prototype for the Su-34 strike/attack platform. The nose shape has led to the nickname 'Platypus'.

Sukhoi Su-27KU/IB

Specification: Sukhoi Su-27IB 'Flanker-?'
Powerplant: Two Lyul'ka AL-31FM (AL-35F) turbofans, each rated at 130.43 kN (29,320 lb st) with afterburning)
Dimensions: wing span 14.70 m (48 ft 2.75 in); length 21.935 m (71 ft 11.5 in) excluding probe; height 5.932 m (19 ft 5.5 in); wing area 46.50 m² (500.54 sq ft)
Weights: maximum take-off 44360 kg (97,795 lb)
Performance: maximum level speed 'clean' at 11000 m (36,090 ft) 1,349 kt (2500 km/h; 1,553 mph) and at sea level 729 kt (1350 km/h; 839 miles); service ceiling 17000 m (55,755 ft); combat radius 809 nm (1500 km; 932 miles) on a hi-hi-hi interception mission with four AAMs
Armament: maximum ordnance 8000 kg (17,636 lb), comprising wide variety of air-to ground and air-to-air stores

Transall C.160

Nouvelle Génération Transalls are distinguished by having an IFR probe above the cockpit. They serve exclusively with the French air force as transports, and in specialised roles.

The Franco-German **Transall C.160** was originally conceived as a Nord Noratlas replacement and was one of the first successful joint European aerospace ventures. Initial procurement comprised 50 **C.160F** aircraft for France and 110 **C.160D** aircraft for West Germany. The first of three prototypes made its maiden flight on 25 February 1963. Production-configured C.160s were delivered from 1967-72. Exports comprised nine **C.160Z**s for South Africa and 20 **C.160T**s (former Luftwaffe examples) for Turkey.

Nouvelle Génération Transalls

The production line was reopened in France in the late 1970s. The AA ordered 29 **C.160NG** (Nouvelle Génération) aircraft which introduced additional fuel capacity, improved avionics and an IFR probe above the cockpit. Maximum payload is 16000 kg (35,275 lb), while 93 troops or 88 paratroops can be accommodated. Ten aircraft were completed with an HDU in the port undercarriage sponson for refuelling tactical aircraft, and five more have provision for this feature so that they can be rapidly reroled as tankers. France's 77 C.160s serve in the transport role with four transport squadrons, various test units and overseas detachments. C.160s also remain active with the Luftwaffe and Turkey. South Africa's C.160Zs were retired in 1993.

Special-duties French variants

France operates six NG aircraft assigned to two forms of special duties. Two have been converted to **C.160GABRIEL** (**C.160G**) Elint and jamming configuration and entered service in 1988. They are distinguished by wingtip pods with blade antennas, five fuselage blade antennas, a blister fairing on each side of the rear fuselage and a retractable ventral dome. Both retain IFR probes and HDUs.

Four **C.160H ASTARTE** Transalls were adapted to carry TACAMO VLF radio transmission equipment (also used by the US Navy's Boeing E-6A Hermes). This takes the form of a long trailing aerial which enables communications with missile-armed nuclear submarines of the Force Océanique Stratégique. The variant entered service in January 1988.

A Luftwaffe C.160 demonstrates its ability to operate from motorways during 1978 Autumn Reforger exercises.

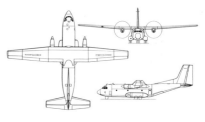

Specification: Transall C.160 (first generation)
Powerplant: two Rolls-Royce Tyne RTy.20 Mk 22 turboprops each rated at 4548 ekW (6,100 ehp)
Dimensions: wing span 40.00 m (131 ft 3 in); length 32.40 m (106 ft 3.5 in); height 11.65 m (38 ft 5 in); wing area 160.10 m² (1,723.36 sq ft)
Weights: empty equipped 28758 kg (63,400 lb); normal take-off 44200 kg (97,443 lb); maximum take-off 49100 kg (108,245 lb); maximum payload 16000 kg (35,273 lb)
Performance: maximum level speed 'clean' at 4500 m (14,765 ft) 289 kt (536 km/h; 333 mph); maximum cruising speed at 5500 m (18,045 ft) 277 kt (513 km/h; 319 mph) and at 8000 m (26,245 ft) 267 kt (495 km/h; 308 mph); maximum rate of climb at sea level 440 m (1,444 ft) per minute; service ceiling 8500 m (27,885 ft); take-off distance to 10.7 m (35 ft) 1100 m (3,609 ft) at MTOW; landing distance from 15 m (50 ft) 640 m (2,100 ft) at normal landing weight; range 2,428 nm (4500 km; 2,796 miles) with an 8000-kg (17,637-lb) payload or 637 nm (1182 km; 734 miles) with a 16000-kg (35,273-lb) payload

Tupolev Tu-16 'Badger'

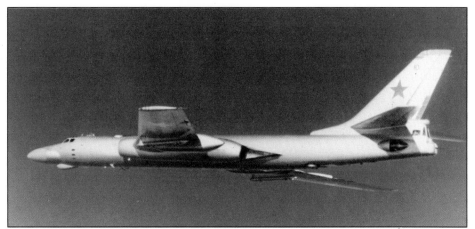

This Tu-16 is a 'Badger-C Mod' missile carrier, with a flat nose radome housing search radar and pylons under the wings for AS-6 'Kingfish' missiles.

The **Tu-16 'Badger'** was developed as a medium bomber to complement the strategic Myasishchev M-4 and Tupolev Tu-95. The prototype first flew on 27 April 1952, and was followed by about 2,000 production **Tu-16s**. Several basic bomber versions were produced, including the **Tu-16A 'Badger-A'** nuclear bomber, the navy's **Tu-16T**, and the SAR-configured **Tu-16K Korvet**. The **Tu-16N** was developed as a tanker using a wingtip-to-wingtip refuelling system. Some 'Badger-As' have been converted with a hose/drogue unit as tankers for probe-equipped bombers.

Missile carriers

The **Tu-16KS-1 'Badger-B'** had a retractable missile guidance radome in the rear of the former bomb bay, and underwing pylons for the carriage of two ASMs. The **Tu-16K-10 'Badger-C'** has a broad, flat nose radome and carries a K-10 (AS-2 'Kipper') missile under the belly. Some were converted to carry the K-26 (AS-6 'Kingfish') missile underwing under the revised reporting name **'Badger-C Mod'**. The **'Badger-G'** was developed as a launch vehicle for the AS-5 'Kelt' missile, with a new target acquisition radar housed below the nose. It was exported to Egypt and Iraq. Some Soviet 'Badger-Gs' were adapted to carry the AS-6 'Kingfish', with new radar under the fuselage.

Elint and reconnaissance variants

The **'Badger-D'** was an Elint conversion based on redundant 'Badger-Cs'. Some 'Badger-As' were converted as **'Badger-E'** dedicated reconnaissance aircraft with a camera/sensor pallet in the former bomb bay. The similar **'Badger-F'** is a dedicated maritime Elint platform, carrying large underwing equipment pods, not usually carried by the similar Elint-tasked **'Badger-K'**. The **Tu-16PP 'Badger-H'** is a dedicated chaff dispenser, while **'Badger-J'** is an active jamming platform with a distinctive ventral canoe fairing. The **'Badger-L'** is an updated maritime Elint/recce platform, with a thimble nose and an extended tailcone.

Licence-production of the 'Badger' took place in China from 1968 until the early 1990s. The air force and navy operate about 100 H-6s, comprising H-6 bombers and H-6D (B-6D) maritime strike aircraft, equipped with two C-601 ASMs.

Tupolev Tu-16 'Badger-F'

Specification: Xian B-6D 'Badger'
Powerplant: two Xian Wopen-8 (Mikulin AM-3M-500) turbojets each rated at 93.16 kN (20,944 lb st) dry
Dimensions: wing span 34.19 m (112 ft 2 in); length 34.80 m (114 ft 2 in); height 10.355 m (33 ft ft 11.75 in); wing area 167.55 m² (1,803.5 sq ft)
Weights: empty equipped 38530 kg (84,944 lb); maximum take-off 75800 kg (167,110 lb)
Performance: (with two C-601 ASMs) maximum cruising speed 424 kt (786 km/h; 488 mph); maximum rate of climb at sea level 1140 m (3,740 ft) per minute; service ceiling 12000 m (39,370 ft); combat radius 971 nm (1800 km; 1,118 miles)
Armament: defensive armament of six guns in dorsal, ventral and tail positions; maximum bombload 9000kg (19,841 lb); primary armament of two 2440-kg (5,379-lb) C-601 ASMs

Tupolev Tu-22 'Blinder'

Russia
Missile carrier/recce platform

The 'Tu-22 Blinder' remains in limited use as a missile carrier, electronic warfare platform and naval surveillance aircraft.

The **Tu-22** was developed to replace the Tu-16, and incorporated supersonic capability to allow penetration of sophisticated defences. The first prototype flew during 1959. The design has a waisted fuselage, engines mounted in pods above the rear fuselage, and the undercarriage carried in trailing-edge pods. The first production version was basically a free-fall bomber and was capable of about Mach 1.5, but had poor endurance and range capabilities. Some may still survive as trainers.

Missile carrier and reconnaissance variant

First appearing in July 1961, the missile-carrying **'Blinder-B'** variant introduced an enlarged undernose radome housing a 'Down Beat' missile guidance radar, and an overnose fairing housing a semi-retractable IFR probe. It carried a single semi-recessed AS-4 'Kitchen' ASM on the centreline, and improved defensive equipment and avionics were fitted in the landing gear pods and wingtip fairings.

Two further variants of the original Tu-22 have been identified, both with the original 'Blinder-A'-type nose radome. The first of these is a dedicated reconnaissance version (possibly designated **Tu-22R**) with camera windows and dielectric panels in the nose and lower fuselage (**'Blinder-C'**). There may be more than one 'Blinder-C' version operational, since different aircraft have different camera and radome configurations, including an Elint-configured sub-variant sometimes reported as **'Blinder-E'**. This has no optical sensors and seems to be used purely in the electronic reconnaissance role. The second variant is the **Tu-22UB** trainer (**'Blinder-D'**) with a raised cockpit for the instructor aft of the normal flight deck.

Between 100 and 136 Tu-22s remain operational with Russia's Long-Range Aviation, and about 20 Tu-22Rs are understood to equip one naval aviation regiment. Tu-22s were also delivered to Libya and Iraq. A handful remain operational in Libya. Used sporadically during the long war with Iran, few of the Iraqi 'Blinders' are likely to have survived air attacks during Desert Storm.

The 'Blinder' was developed to fly the missions of the 'Badger', but with supersonic capability. It is fast disappearing from service, being replaced by the far more capable 'Backfire'.

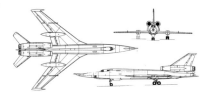

Tupolev Tu-22 'Blinder'

Specification: Tupolev Tu-22 'Blinder-A'
Powerplant: two RKBM (Koliesov) VD-7M turbojets each rated at 156.90 kN (35,273 lb st) with afterburning
Dimensions: wing span 23.75 m (77 ft 11 in); length 40.53 m (132 ft 11.7 in) excluding flight refuelling probe; height 10.67 m (35 ft 0 in)
Weights: basic empty about 40000 kg (88,183 lb); maximum take-off about 83900 kg (184,965 lb)
Performance: maximum level speed 'clean' at 12000 m (39,370 ft) 800 kt (1480 km/h; 920 mph) and at sea level 480 kt (890 km/h; 553 mph); service ceiling 18300 m (60,040 ft); take-off run 2500 m (8,202 ft) at maximum take-off weight; landing run 1600 m (5,249 ft) at normal landing weight; ferry range 3,508 nm (6500 km; 4,039 miles); combat radius 1,673 nm (3100 km; 1,926 miles)
Armament: one radar-directed 23-mm cannon in tail barbette, maximum ordnance 10000 kg (22,046 lb), ('Blinder-B') one AS-4 'Kitchen' missile

Tupolev Tu-22M 'Backfire'

The Tu-22M-3 is the current production 'Backfire' version, identifiable by its wedge-shaped intakes and a new missile targeting and navigation radar in an upturned nose.

Development of the **Tu-22M** was concurrent with the Su-7's VG programme (which led to the Su-17). The two aircraft shared a very similar wing planform, and adopted a similar VG wing. The production **Tu-22M-2 'Backfire-B'** made its first flight in about 1975, and introduced a longer-span wing, a redesigned forward fuselage for four crew and a revised undercarriage, retracting inwards. The tail armament is increased to two remotely-controlled NR-23 23-mm cannons, controlled by the new 'Fan Tail' radar.

Initially, Tu-22Ms were usually seen carrying a single AS-4 'Kitchen' ASM on the centreline, semi-recessed, but today a more usual load seems to be two underwing missiles. In the later **Tu-22M-3 'Backfire-C'**, these bays can accommodate the rotary launchers for the RKV-500B (AS-16 'Kickback') short-range attack missile, used mainly for defence suppression, with two more of these missiles under each wing. Defensive armament is reduced to a single cannon.

'Backfire-C'

The new variant also introduced completely new wedge-type engine intakes, a recontoured upturned nose possibly housing a new attack radar and TFR. 'Backfire-C' is believed to have entered service during 1985, and is the current production version. About 400 'Backfires' of different types are in service, mostly with various Long-Range Aviation strategic regiments, but also including 140 or more with the naval air forces. Production is believed to continue at a rate of 30 per year.

With its wings fully swept back (to 65°), the Tu-22M is capable of a Mach 2 dash at high altitude, and of speeds up to Mach 0.9 at low level. Unrefuelled combat radius of the Tu-22M-2 'Backfire-B' is quoted as 4000 km (2,485 miles), and the radius of action of 'Backfire-C' may be even better. The Tu-22M continues to play a vital role in the defence of the former Soviet Union, albeit now as part of the strategic forces nominally under the central control of the Commonwealth of Independent States. The Tu-22M-3 has also been offered for export and may have been purchased by Iran.

'Backfire-C' forms part of Russia's considerable nuclear delivery force. In addition to strategic missions it has a vital maritime role using nuclear stand-off missiles.

Tupolev Tu-22M-3 'Backfire-C'

Specification: Tupolev Tu-22M-2 'Backfire-B'
Powerplant: probably two KKBM (Kuznetsov) NK-144 each rated at 196.13 kN (44,092 lb st) with afterburning
Dimensions: wing span 34.30 m (112 ft 6.5 in) spread and 23.40 m (76 ft 9.25 in) swept; length 39.60 m (129 ft 11 in); height 10.80 m (35 ft 5.25 in); wing area 170.00 m² (1,829.92 sq ft) spread
Weights: basic empty 54000 kg (119,048 lb); normal take-off 122000 kg (268,959 lb); maximum take-off 130000 kg (286,596 lb)
Performance: maximum level speed 'clean' at 11000 m (36,090 ft) 1,146 kt (2125 km/h; 1,320 mph); service ceiling 18000 m (59,055 ft); ferry range 6,476 nm (12000 km; 7,457 miles); combat radius 2,159 nm (4000 km; 2,486 miles)
Armament: one GSh-23 twin-barrelled 23-mm cannon in tail turret; normal load 12000 kg (26,455 lb), including up to three Kh-22 ASMs

Tupolev Tu-95/142 'Bear'

The 'Bear' variant most seen by the West is the Tu-95RT, a general-purpose long-range maritime patroller. It can also provide mid-course guidance for long-range missiles.

The **Tu-95M 'Bear-A'** entered service in April 1956 as a free-fall nuclear bomber. Surviving 'Bear-As' converted to **Tu-95U** trainers were mostly withdrawn from use during 1991-92. The introduction of stand-off missiles in the 1960s led to the development of various missile-carrying 'Bears'. The **Tu-95K-20 'Bear-B'** was a converted Tu-95M with a broad undernose radome housing 'Crown Drum' guidance radar associated with the AS-3 'Kangaroo' missile. The **Tu-95KM 'Bear-C'** was a new-build equivalent. The **Tu-95K-22 'Bear-G'** is externally similar but has a new 'Down Beat' radar associated with K-22 (NATO AS-4 'Kitchen') missiles.

The **Tu-95RT 'Bear-D'** was developed as a mid-course missile guidance platform, but also has a secondary maritime radar recce role, and serves as an Elint and reconnaissance platform. It has a fixed IFR probe, and a new enlarged chin radome housing 'Big Bulge' search radar. The **Tu-95R 'Bear-E'** was also produced by conversion of Tu-95Ms, and had a reconnaissance pallet in the bomb bay.

Tu-142 'Bear-F/-J' and Tu-95MS 'Bear-H'

The dedicated maritime reconnaissance/ASW **Tu-142 'Bear-F'** incorporated significant improvements, including a strengthened wing, redesigned undercarriage, fuselage plug, uprated NK-12MV engines and redesigned weapons bays. A new ventral radome houses a maritime search radar. The sole export customer is the Indian navy, which received eight Tu-142Ms. The latest **Tu-142M 'Bear-J'** variant fulfils a communications relay role. It has no search radar but has an underfuselage trailing wire antenna pod.

The production line reopened in 1983 to build the new **Tu-95MS-6 'Bear-H'** strategic bomber variant, developed specifically to carry the new RK-55 (AS-15 'Kent') cruise missile. Based on the Tu-142 airframe, 'Bear-H' lacks the fuselage plug of the maritime 'Bears', and introduces a deeper, shorter radome housing an unknown radar. The weapons bay accommodates a rotary launcher for six missiles, and later aircraft (known as **Tu-95MS-16**) carry an additional 10 RK-55s on underwing pylons. It remains in service with Kazakhstan, Russia and Ukraine.

One of the most important 'Bear' variants is the Tu-95MS 'Bear-H'. This is Russia's primary cruise missile carrier.

Tupolev Tu-95 'Bear-D'

Specification: Tupolev Tu-142M 'Bear-F Mod 3'
Powerplant: four KKBM (Kuznetsov) NK-12MV turboprops each rated at 11033 ekW (14,795 ehp)
Dimensions: wing span 51.10 m (167 ft 7.75 in); length 47.50 m (155 ft 10 in) excluding IFR probe and 49.50 m (162 ft 4.8 in) including IFR probe; height 12.12 m (39 ft 9.2 in); wing area 311.10 m² (3,348.76 sq ft)
Weights: empty equipped 86000 kg (189,594 lb); maximum take-off 185000 kg (407,848 lb)
Performance: maximum level speed 'clean' at 7620 m (25,000 ft) 500 kt (925 km/h; 575 mph); cruising speed at optimum altitude 384 kt (711 km/h; 442 mph); climb to 5000 m (16,405 ft) in 13 minutes; service ceiling 12000 m (39,370 ft); combat radius with 11340-kg (25,000-lb) payload 3,454 nm (6400; 3,977 miles)
Armament: twin NR-23 23-mm cannon in tail turret; maximum ordnance 11340 kg (25,000 lb) in two weapon bays for sonobuoys, torpedos, nuclear or conventional depth charges

Tupolev Tu-160 'Blackjack'

Produced to perform the same mission as the B-1, the Tu-160 exhibits a very similar configuration. The materials used for different parts of the airframe's structure are clearly evident.

The **Tu-160 'Blackjack'** is the world's largest bomber, and the heaviest combat aircraft ever built. The Tu-160 was influenced by the Rockwell B-1A, designed to penetrate at high level, relying on performance and a highly sophisticated ECM suite to get through hostile defences. The cancelled B-1 project was subsequently resurrected as the B-1B, which uses low-level subsonic flight and a reduced RCS to penetrate. There was no such cost-cutting exercise in the USSR, and the Tu-160 remains committed to both low-level transonic penetration and high-level supersonic penetration, armed primarily with six RK-55 cruise missiles.

Configuration

The Tu-160's variable-geometry wing and full-span leading-edge slats and trailing-edge double-slotted flaps confer a useful combination of benign low-speed handling and high supersonic speed. Its cockpit is equipped with fighter-type control columns and conventional analog instrument displays, with no MFDs, CRTs and no HUD. The long pointed radome houses a TFR, with a fairing below for the forward-looking TV camera used for visual weapon aiming. A retractable IFR probe endows intercontinental range.

Protracted development

Following a first flight on 19 December 1981, the development programme of the Tu-160 was extremely protracted. Series production took place at Kazan and continued until the termination of further procurement in January 1992. However, limited production may be reinstated, not least to replace aircraft lost to the Ukraine. Even after the aircraft entered service, problems continued to severely restrict operations. These included a shortage of basic flying equipment, problems with the aircraft's K-36A ejection seats and poor reliability of engines and systems.

Eighteen Tu-160s were delivered to the 184th Heavy Bomber Regiment at Priluki from 1987. These were left under Ukrainian command after the break-up of the USSR, but are later to retransfer to Russian control. Four newer aircraft went to Engels, always intended to be the first Tu-160 base.

The Tu-160 has suffered considerable development and operational problems, mostly due to lack of funding.

Tupolev Tu-160 'Blackjack-A'

Specification: Tupolev Tu-160 'Blackjack-A'
Powerplant: four SSPE Trud (Kuznetsov) NK-321 turbojets each rated at 137.20 kN (30,843 lb st) dry and 245.16 kN (55,115 lb st) with afterburning
Dimensions: wing span 55.70 m (182 ft 9 in) spread and 35.60 m (116 ft 9.75 in) swept; length 54.10 m (177 ft 6 in); height 13.10 m (43 ft 0 in); wing area 360.00 m² (3,875.13 sq ft)
Weights: empty equipped 118000 kg (260,140 lb); normal take-off 267600 kg (589,947 lb); maximum take-off 275000 kg (606,261 lb)
Performance: maximum level speed 'clean' at 11000 m (36,090 ft) 1,079 kt (2000 km/h; 1,243 mph); long-range cruising speed at optimum altitude 460 kt (850 km/h; 528 mph); range 7,555 nm (14000 km; 8,699 miles)
Armament: maximum ordnance about 16330 kg (36,000 lb); two tandem fuselage weapons bays, each normally equipped with a rotary carousel for six RK-55 (AS-15 'Kent') cruise missiles (with 200-kT warhead and range in excess of 3000 km), 12 Kh-15P (AS-16 'Kickback') 'SRAMskis' or free-fall bombs

125

Westland Lynx

The Lynx HAS.Mk 3 serves aboard Royal Navy ships on a variety of duties. The major tasking is for ASW, but anti-ship attacks and utility duties are also undertaken.

Launched under the Anglo-French helicopter agreement of February 1967, the **Westland Lynx** is an extremely versatile and agile helicopter with digital flight controls and a four-bladed semi-rigid main rotor. The first prototype flew on 21 March 1971. The production **Lynx HAS.Mk 2** undertook a range of shipboard missions including ASW, SAR, ASV search and strike, reconnaissance, troop transport, and VertRep duties. The **HAS.Mk 3** introduced Gem 41-1 turboshafts, and subsequent upgrades included the **HAS.Mk 3S** with secure speech facility, the **HAS.Mk 3ICE** for use aboard the Antarctic survey vessel *Endurance*, the **HAS.Mk 3CTS** with a new central tactical system and the **HAS.Mk 3GM** with improved cooling, IR jammers and ALQ-167 ECM pods. Foreign customers for the first-generation naval Lynx were Argentina, Brazil, Denmark, France, Germany, the Netherlands, Nigeria and Norway.

Most surviving AAC, RN and Aéronavale Lynxes are receiving new composite rotor blades with anhedralled and swept high-speed tips, developed through the BERP programme. The **HAS.Mk 8** has these and a nose-mounted thermal imager turret, a rear-mounted MAD, Orange Crop ESM and a Yellow Veil ECM jamming pod. Many Lynx Mk 8 features are incorporated in the export **Super Lynx**, ordered by Brazil, South Korea and Portugal.

Army Lynxes

The second production model was the **Lynx AH.Mk 1** battlefield helicopter, first flying in February 1977. It has skid landing gear and can carry 12 troops or 907 kg (2,000 lb) of internal freight, or a wide assortment of weapons including eight TOW anti-tank missiles aimed using a roof-mounted sight. The **Lynx AH.Mk 7** has the direction of tail rotor rotation reversed, uprated Gem 41 engines and a box-like IR exhaust shroud. The **Lynx AH.Mk 9** incorporates all the AH.Mk 7 modifications and features a nosewheel undercarriage, and exhaust diffusers. The first new-build example flew on 20 July 1990. Defence economies precluded the provision of TOW ATGM capability, although this may follow if funds permit, using a modified weapons pylon.

Lynx AH.Mk 7s form the backbone of the British Army's anti-tank helicopter force, each equipped with eight TOWs.

Westland Lynx AH.Mk 9

Specification: Westland Lynx AH.Mk 7
Powerplant: two Rolls-Royce Gem 42-1 turboshafts each rated at 846 kW (1,135 shp)
Dimensions: main rotor diameter 12.80 m (42 ft 0 in); length overall, rotors turning 49 ft 9 in (15.16 m); height overall 12 ft 0 in (3.73 m) with rotors stationary; main rotor disc area 128.71 m² (1,385.44 sq ft)
Weights: operating empty 3072 kg (6,772 lb) in the anti-tank role; maximum take-off 4876 kg (10,750 lb); maximum payload 3,000 lb (1361 kg)
Performance: maximum continuous cruising speed at optimum altitude 138 kt (256 km/h; 159 mph); maximum rate of climb at sea level 756 m (2,480 ft) per minute; hovering ceiling 3230 m (10,600 ft); combat radius 25 nm (46 km; 29 miles) for a 2-hour patrol on an anti-tank mission
Armament maximum ordnance about 550 kg (1,210 lb), including eight TOW missiles

Westland Sea King

For assault transport, the Royal Navy uses the Sea King HC.Mk 4, known as the Commando for export customers. It is distinguished by its lack of fuselage sponsons.

Four Sikorsky-built S-61s were shipped to Westland in 1966 as pattern aircraft for licence-production of the **Sea King HAS.Mk 1**. Flying on 7 May 1969, it had British avionics and ASW systems, including search radar, sonar, Doppler and bathythermographic equipment, and AFCS. Weapons included Mk 44 torpedoes, Mk 11 depth charges or a WE177 nuclear depth bomb. The 56 RN HAS.Mk 1s were followed by 21 **HAS.Mk 2**s (and 37 conversions) with uprated engines, six-bladed tail rotors and intake deflector/filters.

The **Sea King HAS.Mk 5** (85 produced, some by conversion) featured Sea Searcher radar, improved processing and Orange Crop ESM. The **HAS.Mk 6** has an integrated sonar processor, deep water dipping sonar, and Orange Reaper ESM. Five new-build HAS.Mk 6s followed the prototype (a converted HAS.Mk 5) and 69 conversion kits have been supplied to RNAY Fleetlands. The similar **Advanced Sea King** was developed specifically for export, India buying 12 Sea Eagle-compatible **Sea King Mk 42B**s and six SAR **Sea King Mk 42C**s. The withdrawal of HMS *Ark Royal* left the Royal Navy without shipborne AEW cover. The **Sea King AEW.Mk 2** has a Searchwater radar with an inflatable radome. Three similar AEW radar systems were sold to Spain for conversion of Spanish navy SH-3H Sea Kings.

Nineteen dedicated **Sea King HAR.Mk 3** SAR aircraft were delivered to the RAF and six more were ordered in February 1992. These are to be equipped to an even higher standard, with a new radar, a new AFCS and improved nav systems and radios. ASW and SAR Sea Kings were exported to Australia, Belgium, Egypt, Germany, India, Norway, and Pakistan.

Commando

Development of an assault/tactical transport Sea King began in mid-1971, resulting in the **Westland Commando**. No interest was initially expressed by the UK, but Egypt and Qatar placed orders. RN interest began in 1978, as a replacement for the Wessex. Forty-one **Commando HC.Mk 4**s were procured for the RN. Two similar **Sea King Mk 4X**s were built for the RAE as test aircraft.

The Sea King AEW.Mk 2 provides low-cost AEW cover for the fleet. The radar is housed in an inflatable radome.

Westland Sea King HAS.Mk 6

Specification: Westland Advanced Sea King HAS.Mk 6
Powerplant: two Rolls-Royce Gnome H.1400-1T turboshafts each rated at 1238 kW (1,660 shp) for take-off and 1092 kW (1,465 shp) for continuous running
Dimensions: main rotor diameter 18.90 m (62 ft 0 in); length overall, rotors turning 22.15 m (72 ft 8 in), fuselage 17.02 m (55 ft 10 in); height overall 5.13 m (16 ft 10 in) with rotors turning; main rotor disc area 280.47 m² (3,019.07 sq ft)
Weights: empty equipped 7428 kg (16,377 lb) – ASW role, maximum take-off 9752 kg (21,500 lb)
Performance: maximum cruising speed at sea level 110 kt (204 km/h; 126 mph); maximum rate of climb at sea level 619 m (2,030 ft) per minute; service ceiling 1220 m (4,000 ft) with one engine out; range 800 nm (1482 km; 921 miles)
Armament: maximum ordnance 1134 kg (2,500 lb), comprising torpedoes, depth charges and ASMs

Glossary

AA	Armée de l'Air	EWO	Electronic Warfare Officer	ODR	Overland Downlook Radar
AAM	Air-to-Air Missile	FAC	Forward Air Control	OTH	Over The Horizon
ACC	Air Combat Command	FBW	Fly By Wire	PGM	Precision-Guided Munitions
AEW	Airborne Early Warning	FCS	Flight Control System	PLA	People's Liberation Army
AEW&C	Airborne Early Warning and Command	FIS	Fighter Intercept Squadron	PLSS	Precision Location Strike System
		FLIR	Forward-Looking Infra-Red	PNVS	Pilot's Night Vision System
AFB	Air Force Base	FSD	Full-Scale Development	PVO	Protivovosduzhnaya Oborona – (Russian) Air Defence
AFCS	Automated Flight Control System	GCI	Ground-Controlled Interception		
AFRES	Air Force REServe	GPS	Global Positioning System	RAF	Royal Air Force
AFV	Armoured Fighting Vehicle	HARM	High-speed Anti-Radiation Missile	RAM	Radar Absorbent Material
AGM	Air-to-Ground Missile	HDD	Head-Down Display	RANSAC	RANge Surveillance AirCraft
ALAT	Aviation Légère de l'Armée de Terre	HDU	Hose-Drum Unit	RAST	Recover, Assist, Secure, Traverse
AMC	Air Mobility Command	HIRSS	Helicopter Infra-Red Suppression System	RCS	Radar Cross-Section
AMP	Avionics Modernisation Programme			RDT & E	Research, Development, Test and Evaluation
AMRAAM	Advanced Medium-Range Air-to-Air Missile	HMMWV	High-Mobility Multi-purpose Wheeled Vehicle	RHAWS	Radar Homing And Warning System
ANG	Air National Guard	HOT	Haut subsonique Optiquement Téléguidé tiré d'un Tube	RN	Royal Navy
AoA	Angle of Attack			RNAY	Royal Naval Aircraft Yard
APC	Armoured Personnel Carrier	HOTAS	Hands On Throttle And Stick	RNZAF	Royal New Zealand Air Force
APU	Auxiliary Power Unit	HUD	Head-Up Display	RSAF	Royal Saudi Air Force
ARIA	Advanced Range Instrumentation Aircraft	IAS	Indicated Air Speed	RWR	Radar Warning Receiver
		ICBM	InterContinental Ballistic Missile	SAAF	South African Air Force
ARM	Anti-Radiation Missile	IDF/AF	Israeli Defence Force/Air Force	SAC	Strategic Air Command
ARNG	ARmy National Guard	IFF	Identification, Friend or Foe	SAM	Surface-to-Air Missile
ASM	Air-to-Surface Missile	IFR	In-Flight Refuelling	SAOEU	Strike/Attack Operational Evaluation Unit
AShM	Anti-Ship Missile	IIR	Imaging Infra-Red		
ASMP	Air-Sol Moyenne Portée	IOC	Initial Operating Capability	SAR	Search And Rescue
ASW	Anti-Submarine Warfare	INAS	Inertial Navigation and Attack System	SARH	Semi-Active Radar Homing
ATARS	Advanced Tactical Air Reconnaissance System			SEAD	Suppression of Enemy Air Defences
		INS	Inertial Navigation System	SEAL	SEa, Air, Land
ATF	Advanced Tactical Fighter	IR	Infra-Red	Sigint	Signals intelligence
ATGM	Anti-Tank Guided Missile	IRST	Infra-Red Search and Track	SLAM	Stand-off Land Attack Missile
ATM	Anti-Tank Missile	JASDF	Japan Air Self-Defence Force	SLAR	Side-Looking Airborne Radar
AWACS	Airborne Warning and Control System	JGSDF	Japan Ground Self-Defence Force	SLEP	Service Life Extension Programme
BERP	British Experimental Rotor Programme	JMSDF	Japan Maritime Self-Defence Force	SRAM	Short-Range Attack Missile
BVR	Beyond Visual Range	JPATS	Joint Primary Aircraft Trainer System	STAR	Surface-To-Air Recovery
C³I	Command, Control, Communications and Intelligence	J-STARS	Joint Surveillance Target Attack Radar System	STO	Short Take-Off
				STOL	Short Take-Off and Landing
CAP	Combat Air Patrol	JTIDS	Joint Tactical Information Distribution System	STOVL	Short Take-Off and Vertical Landing
CBU	Cluster Bomb Unit			TACAMO	TAke Charge And Move Out
CAF	Canadian Armed Forces	LAMPS	Light Airborne Multi-Purpose System	TACAN	TACtical Air Navigation
CAS	Corrected Air Speed	LANTIRN	Low-Altitude, Navigation and Targeting, Infra-Red, for Night	TADS	Target Acquisition and Designation Sight
CFT	Conformal Fuel Tank				
CIA	Central Intelligence Agency	LAPES	Low-Altitude Parachute Extraction System	TCS	Television Camera System
CIS	Commonwealth of Independent States			Telint	Telemetry intelligence
		LASTE	Low-Altitude Safety and Targeting Enhancement	TFR	Terrain-Following Radar
CRT	Cathode Ray Tube			TINS	Tactical Inertial Navigation System
COD	Carrier Onboard Delivery	LERX	Leading-Edge Root eXtensions	TJS	Tactical Jamming System
COIN	COunter-INsurgency	LGB	Laser-Guided Bomb	TRAM	Target Recognition and Attack Multisensor
CVW	Carrier air wing	LLLTV	Low-Light-Level TeleVision		
CW	Continuous Wave	LOROP	LOng-Range Oblique Photography	TOW	Tube-launched, Optically-sighted, Wire-guided
DACT	Dissimilar Air CombaT	LRMTS	Laser Rangefinder and Marked Target Seeker		
DARPA	Defense Advanced Research Projects Agency	MAC	Military Airlift Command	UK	United Kingdom
		MAD	Magnetic Anomaly Detection	UN	United Nations
DIANE	Digital, Integrated Attack and Navigation Equipment	MBT	Main Battle Tank	USAF	US Air Force
		MFD	Multi-Function Display	USCG	US Coast Guard
DLIR	Downward-Looking Infra-Red	MIRLS	Miniaturised Infra-Red Landing System	USMC	US Marine Corps
DOSAAF	Dobrovol'noe obshchestvo sodietstvija Armii, Aviatii i Flotu – (Russian) Vluntary Society for the Support of the Army, Aviation and Fleet - paramilitary youth sport-flying organisation			USN	US Navy
		MMW	MilliMetric Wave	USSR	Union of Soviet Socialist Republics
		MPLH	Multi-Purpose Light Helicopter	VertRep	Vertical Replenishment
		MSIP	Multi-Stage Improvement Programme	VG	Variable Geometry
				VIP	Very Important Person
		MTOW	Maximum Take-Off Weight	VLF	Very Low Frequency
		NACES	Naval Aircrew Common Ejection System	VTA	Voenno-Transportnaya Aviatsiya – (Russian) Military Transport Command
ECR	Electronic Combat Reconnaissance				
ECM	Electronic CounterMeasures	NASA	National Aeronautics and Space Administration		
Elint	Electric intelligence			VTO	Vertical Take-Off
EO	Electro-optical	NATO	North Atlantic Treaty Organisation	VVS	Voenno-Vozdushnye Sili – (Russian) Air Forces
ESM	Electronic Surveillance Measures	NBC	Nuclear, Biological, Chemical		
ESSS	External Stores Support System	NVG	Night Vision Goggles	WSO	Weapons System Operator
EVS	Electro-optical Viewing System	OCU	Operational Conversion Unit		
EW	Electronic Warfare				

Picture acknowledgments

6: Aermacchi (two). 7: Aero. 8: Agusta (two). 9: AIDC (two). 10: AMX International (two). 11: CNAIEC. 12: Indian Air Force. 14: Atlas (two). 15: Beech (two). 16: Bell (two). 17: Bell (two). 18: Bell (two). 19: Boeing (two). 20: David Donald, MAP. 21: Lockheed. 22: NASA. 23: Boeing. 24: Grumman (two). 25: Boeing/Sikorsky. 26: Boeing Helicopters, US Navy. 27: Boeing Helicopters, UK MoD. 28: Boeing Helicopters (two). 29: British Aerospace, UK MoD. 30: British Aerospace, Rolls-Royce. 31: UK MoD, British Aerospace. 32: British Aerospace (two). 33: ENAER, CASA. 34: Peter Steinemann, Sabreliner Corp. 35: CNAIEC. 36: Dassault. 37: Dassault. 38: Dassault (two). 39: Dassault, GIFAS. 40: Dassault (two). 41: Dassault (two). 42: Dassault, Luftwaffe. 43: Westland, Agusta. 44: Northrop, EMBRAER. 45: Eurocopter. 46: Eurocopter, MATRA. 47: Eurocopter (two). 48: Eurocopter, British Aerospace. 49: British Aerospace (two). 51: US Air Force, RAAF. 52: Grumman (two). 53: Grumman, US Navy. 54: Grumman, US Navy. 55: US Navy (two). 56: US Navy (two). 57: US Air Force, Grumman. 58: IAI (two). 59: US Navy (two). 60: Royal Norwegian Air Force (two). 61: Sergei Skrynnikov (two). 62: Kawasaki (two). 63: Lockheed (two). 64: Lockheed, UK MoD. 65: Rockwell. 66: US Air Force (two). 67: Lockheed (two). 68: Lockheed (two). 69: Aeritalia (two). 70: Lockheed (two). 71: Lockheed (two). 72: US Navy (two). 73: Lockheed (two). 74: Lockheed (two). 75: Peter Steinemann. 76: McDonnell Douglas (two). 77: McDonnell Douglas. 78: IAI, Frazer-Nash. 79: McDonnell Douglas (two). 80: McDonnell Douglas (two). 81: RAAF, McDonnell Douglas. 82: McDonnell Douglas (two). 83: McDonnell Douglas (two). 84: McDonnell Douglas (two). 85: McDonnell Douglas (two). 86: McDonnell Douglas. 87: Ryszard Jaxa-Malachowski. 88: US Navy. 91: Indian Air Force. 92: David Donald, ITAR-TASS. 93: ITAR-TASS, Mikoyan. 94: Indian Air Force. 96: John Fricker. 97: MAP. 98: CNAIEC. 99: Vought, Northrop. 101: IAI, Northrop. 102: British Aerospace (two). 103: British Aerospace. 104: Pilatus, KLu. 105: Rockwell (two). 106: Saab (two). 107: Saab (two). 108: Saab (two). 109: British Aerospace, Aérospatiale. 110: CNAIEC (two). 111: Agusta, Sikorsky. 112: Sikorsky. 113: Sikorsky, US Air Force. 114: Sikorsky, US Navy. 120: Transall. 121: US Navy, CNAIEC. 124: UK MoD, US Air Force. 125: Sebastian Zacharias. 126: Westland, Army Air Corps. 127: Westland (two).

INDEX
Aircraft listed by manufacturer, name and designation

Glossary 128